Related Books of Interest

Developing Quality Technical Information, Second Edition

By Gretchen Hargis, Michelle Carey, Ann Kitty Hernandez, Polly Hughes, Deirdre Longo, Shannon Rouiller, and Elizabeth Wade

ISBN: 0-13-47749-8

Direct from IBM's own documentation experts, this is the definitive guide to developing outstanding technical documentation—for the Web and for print. Using extensive before-and-after examples, illustrations, and checklists, the authors show exactly how to create documentation that's easy to find, understand, and use. This edition includes extensive new coverage of topic-based information, simplifying search and retrievability, internationalization, visual effectiveness, and much more.

The IBM Style Guide

By Francis DeRespinis, Peter Hayward, Jana Jenkins, Amy Laird, Leslie McDonald, and Eric Radzinski

ISBN: 0-13-210130-0

The IBM Style Guide distills IBM wisdom for developing superior content: Information that is consistent, clear, concise, consumable, reusable, and easy to translate. Written by a team of senior IBM editors, this book helps any organization improve and standardize content across authors, delivery mechanisms, and geographic locations. *The IBM Style Guide* can help any organization or individual create and manage content more effectively.

Related Books of Interest

Enterprise Master Data Management

An SOA Approach to Managing Core Information

By Allen Dreibelbis, Eberhard Hechler, Ivan Milman, Martin Oberhofer, Paul Van Run, and Dan Wolfson

ISBN: 0-13-236625-8

The Only Complete Technical Primer for MDM Planners, Architects, and Implementers

Enterprise Master Data Management provides an authoritative, vendor-independent MDM technical reference for practitioners: architects, technical analysts, consultants, solution designers, and senior IT decision makers. Written by the IBM® data management innovators who are pioneering MDM, this book systematically introduces MDM's key concepts and technical themes, explains its business case, and illuminates how it interrelates with and enables SOA.

Drawing on their experience with cutting-edge projects, the authors introduce MDM patterns, blueprints, solutions, and best practices published nowhere else—everything you need to establish a consistent, manageable set of master data, and use it for competitive advantage.

The Art of Enterprise Information Architecture

A Systems-Based Approach for Unlocking Business Insight

By Mario Godinez, Eberhard Hechler, Klaus Koenig, Steve Lockwood, Martin Oberhofer, and Michael Schroeck

ISBN: 0-13-703571-3

Architecture for the Intelligent Enterprise: Powerful New Ways to Maximize the Real-time Value of Information

In this book, a team of IBM's leading information management experts guide you on a journey that will take you from where you are today toward becoming an "Intelligent Enterprise."

Drawing on their extensive experience working with enterprise clients, the authors present a new, information-centric approach to architecture and powerful new models that will benefit any organization. Using these strategies and models, companies can systematically unlock the business value of information by delivering actionable, real-time information in context to enable better decision-making throughout the enterprise—from the "shop floor" to the "top floor."

Related Books of Interest

Search Engine Marketing, Inc

By Mike Moran and Bill Hunt

ISBN: 0-13-606868-5

The #1 Step-by-Step Guide to Search Marketing Success...Now Completely Updated with New Techniques, Tools, Best Practices, and Value-Packed Bonus DVD!

In this book, two world-class experts present today's best practices, step-by-step techniques, and hard-won tips for using search engine marketing to achieve your sales and marketing goals, whatever they are. Mike Moran and Bill Hunt thoroughly cover both the business and technical aspects of contemporary search engine marketing, walking beginners through all the basics while providing reliable, up-to-the-minute insights for experienced professionals.

Thoroughly updated to fully reflect today's latest search engine marketing opportunities, this book guides you through profiting from social media marketing, site search, advanced keyword tools, hybrid paid search auctions, and much more.

Listen to the author's podcast at:
ibmpressbooks.com/podcasts

Audience, Relevance, and Search

Targeting Web Audiences with Relevant Content

Mathewson, Donatone, Fishel

ISBN: 0-13-700420-6

The Social Factor

Innovate, Ignite, and Win Through Mass Collaboration and Social Networking

Azua

ISBN: 0-13-701890-8

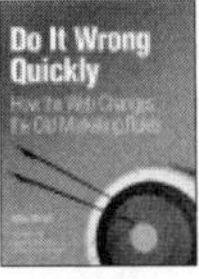

Do it Wrong Quickly

How the Web Changes the Old Marketing Rules

Moran

ISBN: 0-13-225596-0

Get Bold

Using Social Media to Create a New Type of Social Business

Carter

ISBN: 0-13-261831-1

Mining the Talk

Unlocking the Business Value in Unstructured Information

Spangler, Kreulen

ISBN: 0-13-233953-6

Sign up for the monthly IBM Press newsletter at ibmpressbooks/newsletters

DITA Best Practices

DITA Best Practices

A Roadmap for Writing, Editing, and Architecting in DITA

Laura Bellamy

Michelle Carey

Jenifer Schlotfeldt

IBM Press
Pearson plc
Upper Saddle River, NJ • Boston • Indianapolis • San Francisco
New York • Toronto • Montreal • London • Munich • Paris • Madrid
Cape Town • Sydney • Tokyo • Singapore • Mexico City

ibmpressbooks.com

IBM Press Program Managers: Steven M. Stansel, Ellice Uffer
Cover design: IBM Corporation

Editor in Chief: Bernard Goodwin
Marketing Manager: Stephane Nakib
Publicist: Heather Fox
Acquisitions Editor: Bernard Goodwin
Managing Editor: Kristy Hart
Designer: Alan Clements
Senior Project Editor: Lori Lyons
Copy Editor: Apostrophe Editing Services
Proofreader: Williams Woods Publishing Services
Manufacturing Buyer: Dan Uhrig

Published by Pearson plc

Publishing as IBM Press

IBM Press offers excellent discounts on this book when ordered in quantity for bulk purchases or special sales, which may include electronic versions and/or custom covers and content particular to your business, training goals, marketing focus, and branding interests. For more information, please contact:

U.S. Corporate and Government Sales
1-800-382-3419
corpsales@pearsontechgroup.com

For sales outside the U.S., please contact:

International Sales
international@pearson.com

The following terms are trademarks or registered trademarks of International Business Machines Corporation in the United States, other countries, or both: IBM, IBM Press, Lotus and Notes.

UNIX is a registered trademark of The Open Group in the United States and other countries or both.

Linux is a registered trademark of Linus Torvalds in the United States, other countries, or both.

Microsoft, Windows, SharePoint and the Windows logo are trademarks of Microsoft Corporation in the United States, other countries, or both.

Java and all Java-based trademarks and logos are trademarks or registered trademarks of Oracle and/or its affiliates

Adobe, the Adobe logo, FrameMaker, InDesign, PhotoShop and Illustrator are either registered trademarks or trademarks of Adobe Systems Incorporated in the United States, and/or other countries.

Other company, product, or service names may be trademarks or service marks of others.

Library of Congress Cataloging-in-Publication data is on file.

Pearson Education, Inc
Rights and Contracts Department
501 Boylston Street, Suite 900
Boston, MA 02116
Fax (617) 671-3447

First printing October 2011

ISBN-13: 978-0-13-248052-9
ISBN-10: 0-13-248052-2

Contents

Acknowledgments

A single page could never express our gratitude to the family, friends, and colleagues who contributed to this effort. But we're not quitters, so we'll attempt to thank everyone who made this book possible.

We appreciate the camaraderie and support from the DITA community. The user groups, forums, and conferences have helped to shape our DITA knowledge. We want our peers and colleagues to understand how valuable they have been.

We are fortunate to have such leaders in the DITA community as reviewers: Thank you to Don Day, chair of the OASIS DITA Technical Committee, for taking the time to be one of our reviewers. Your DITA knowledge is vast and we appreciate the help. Thank you to Yas Etessam for lending us your detailed technical knowledge of DITA and your experience leading DITA implementations at such diverse companies. Amber Swope brought a more complete view for converting content to DITA. Heather Crognale is the best kind of friend, the one who reviews your manuscript and still sends you a Christmas card. Because of her years of experience as a DITA editor and her eagle eye as a copy editor, she has helped others to become DITA editors. Evelyn Eldridge is already a fan of editing in DITA, and her contribution to this book will spread the enthusiasm.

Thank you to Rob Lee for designing the icons. Without his contribution, we would never have known that teal is the new *in* color.

We would also like to thank Janine Trakhtenberg and Just Systems whose XMetaL authoring tool we used to produce the sample DITA files and book examples.

Of course, no book would be worth the reading if that book weren't given a good working over by editors. Good editors simply make writers look better. So thank you to Shannon Rouiller, who provided expert advice on short descriptions, topic-based writing, and other chapters. Shannon's thoughtful and kind advice helped us turn rough drafts into coherent information. Thanks to Elizabeth Wilde who showed us a new direction for the topic-based writing chapters and helped us dig deeper to find better examples, better tone, and better organization. And thanks to Marianne White, who helped us improve the consistency of our terminology, improve our examples, catch all those embarrassing nits, and improve the flow of the chapters.

Finally, a special thanks to our families and friends. It'll take a lot more than a few words to make up for all the late nights, the missed weekends together, and the constant dinner conversation about the default behavior of related links. Throughout this effort, our families and friends have become unwitting DITA experts. Despite the fact that knowing DITA is a very marketable skill, we realize that they put up with quite a lot. Their support, encouragement, and steady supply of caffeine made all the difference.

About the Authors

Laura Bellamy is an Information Architect at VMware, Inc. and a technical communications instructor at University of California Santa Cruz Extension. Laura has been a long-time DITA champion, working at IBM during the adoption and proliferation of DITA. Throughout her career, she has worked on many facets of DITA implementation and now dreams in XML.

Michelle Carey is a technical editor at IBM and a technical commu-nications instructor at University of California Santa Cruz Extension. Michelle has taught IBM teams and users' groups about best practices for authoring in DITA, topic-based writing, writing for translation, editing user interfaces, and writing effective error messages. She is also a coauthor of the book *Developing Quality Technical Information*. Michelle loves to ride motorcycles and mountain bikes, herd cats, and diagram sentences.

Jenifer Schlotfeldt is a project leader, information developer, and technical leader at IBM and a technical communications instructor at the University of California Santa Cruz Extension. She has been authoring, testing, and teaching DITA since 2003. She has converted documentation to DITA, authored new content in DITA, contributed to new DITA specializations, and created many training materials for different facets of DITA authoring.

Introduction

So, finally, you've decided to write your content in DITA.

First, congratulations! For most companies or organizations, deciding whether to adopt Darwin Information Typing Architecture (DITA) as an XML authoring methodology is an arduous journey. We're strong supporters of the DITA standard and are confident that your investment to move to DITA can improve the technical information that you create for your products, services, or technologies.

Second, now what are you going to do? How do you start implementing DITA? What do you need to know before you start writing? What best practices have many of us in the technical writing community established? What are the gotchas and pitfalls of authoring in DITA? What's the best way to learn about DITA?

Don't panic: We've helped to educate many teams in our own companies and communities and promise you that learning how to semantically tag content in XML is not the hard part. Learning how to best use (or not use) the DITA elements, DITA maps, and topic types to fit your content and your organization is the real challenge.

The default DITA standard includes more than 400 elements, and those elements contain attributes. You'll need to consider these questions before you get started:

- Should you use all of the DITA elements and attributes?
- Which features do you need to understand to get going?
- Which features can you wait to implement after you've spent more time working with DITA?
- What are the guidelines that you should follow?

This book is for users who have made the decision to use DITA and are looking for advice from experienced DITA authors, editors, and information architects about how to write effective technical information in DITA.

We decided to write this book to fill an information gap in the available DITA information. Some DITA books and educational courses tell you what each DITA element represents and what it is used for. However, they don't always tell you how to best create effective content for these elements or how to organize that content.

For example, current DITA education defines what a <shortdesc> element is, but that education doesn't show you how to write effective text to go in that element.

We've spent years evaluating, testing, and writing best practices for technical writers at our companies. Our recommendations go beyond defining elements: We'll show you how and when to use an element, how to write effective text for that element, and even when not to use specific elements and attributes.

One consistent question that we hear at conferences and community discussions is, "How did you agree on best practices and write those guidelines?" The answer is years and years of trial and error until we refined our guidelines to a set of standards that help us to create industry-leading technical information.

For example, we've presented our short description best practices at DITA conferences and user group forums and meetings. At one presentation, a colleague said, "I wish someone would sell us these best practices so we don't have to spend months writing our own guidelines." Good idea!

Writers, editors, information architects, and even managers will find helpful guidelines and best practices for writing, organizing, and editing DITA content, and converting non-DITA content to DITA topics:

- Technical writers will learn how topic-based writing in DITA can help them create more effective information.
- Editors will learn about new ways to ensure the quality of the information.
- Information and XML architects will get practical advice about which DITA elements and features to implement.
- Project managers will find roadmaps and checklists to help them coordinate the conversion to DITA.
- Technical writing managers will find information about roles and resources required for converting content to DITA.

We created these examples and best practices by using out-of-the-box DITA authoring and processing tools. We used DITA 1.1 in the XMetaL authoring tool and produced PDF and HTML output by using the DITA Open Toolkit. In some cases, we highlighted features that are unique to DITA 1.2. You, too, can create effective topic-based information by using the default settings in DITA.

You can find a number of resources to help you learn DITA basics, and the DITA community is a wonderfully supportive group.

In some chapters, we refer to two other books that cover similar topics: *The IBM Style Guide: Conventions for Writers and Editors* by DeRespinis et al. (ISBN: 9780132101301), and *Developing Quality Technical Information: A Handbook for Writers and Editors* by Hargis et al. (ISBN: 9780137034574), both published by IBM Press (Pearson Education, Inc.).

We hope that you find these guidelines helpful. Now, let's have some fun with DITA!

PART I

Writing in DITA

One of the revelations of minimalism is the task-oriented topic that focuses on the goals of the user. To make writing that topic easier, DITA was born. Darwin Information Typing Architecture (DITA) was designed specifically for writing, organizing, and linking topic-based content. The benefits of DITA aren't limited to customers who read product information. DITA has many benefits for those who write and edit technical information before it goes to customers.

For users of your information, well-written, topic-based information improves retrievability, navigation, and usability. For writing teams, effective topic-based writing provides more opportunities for reuse, quick reorganization of information, easier file management, and more flexible linking.

To make your move to DITA a little easier, Part I provides guidance for writing task, concept, and reference topics. You'll learn how to write effective content for many of the common and sometimes challenging DITA XML elements inside those topics.

Part I also describes how to write effective short descriptions for the <shortdesc> element, which is perhaps the most challenging element to write for—so challenging, in fact, that Chapter 5 is dedicated to it.

CHAPTER 1

Topic-Based Writing in DITA

What's a topic other than a conversational piece? In technical information, a *topic*, which is sometimes called an *article*, has a title and some content. A topic has just enough content to make sense by itself but not so much content that it covers more than one procedure, one concept, or one type of reference information.

Although a topic should be self-contained, it shouldn't live alone. For information delivered with most technical products, services, or technologies, a topic needs a home in a larger, organized collection of topics. That collection can then be packaged and delivered in an output format such as HTML web pages, online help, or a PDF manual.

Darwin Information Typing Architecture (DITA) is specifically designed to support topic-based writing. With its modular-based architecture, semantic XML elements, and powerful linking features, DITA can help you create and maintain topic-based, or componentized, technical information.

Writing topic-based information has advantages for both the users of your products and for your writing teams. However, separating your information into different topic types isn't enough. To ensure that your information meets the needs of your users, you need to understand how to write minimalist, task-oriented information.

Books, Topics, and Webs of Information

Books are great for some things, such as mystery novels, complex scientific concepts, and philosophical ponderings. However, books aren't the best vehicle for delivering targeted technical content to users who have real-world tasks to complete.

Information organized in a narrative book model typically has these characteristics:

- Beginnings, middles, and endings intended to be read linearly from beginning to end
- Chapters or sections that often mix task, conceptual, and reference information

Although the narrative book model works well for novels and some nonfiction books, it's not well suited for technical information that's delivered with a product. When it's time to adjust the valves on the motorcycle, users don't want to read a novel. They want to open the motorcycle manual, find that one specific task, and move on.

A topic is a self-contained unit of information. An effective topic covers only one subject. Each topic is long enough to make sense on its own, but short enough to stick to one point without expanding into other subjects.

A topic typically answers one of these questions:

- How do I do it?
- What is it?
- What is the process?
- How do I fix it?

Because most users need only a small amount of information at a time, you should create information that can answer specific questions discretely without requiring extensive reading across large amounts of interconnected content.

Separating information into discrete topics helps you to:

- Design new information more easily and consistently
- Eliminate unimportant or redundant information
- Identify common or reusable topics or pieces of content in topics

All topics, regardless of their purpose, have the following characteristics:

- Meaningful titles
- Ability to stand alone from other content
- Logical organization
- Links to other topics that contain related information

You organize and link topics to create a coherent web of information, as shown in Figure 1.1. Eventually, the output of this collection of topics might be an online help system, a PDF manual, or a website. Although a web of topics might seem chaotic at first, DITA can help you organize and link topics so that users don't get lost.

By writing your information in discrete topics, organizing those topics into logical collections, and then linking related topics to each other, you can create a web of information that is easy to navigate, easy to understand, and easy to consume.

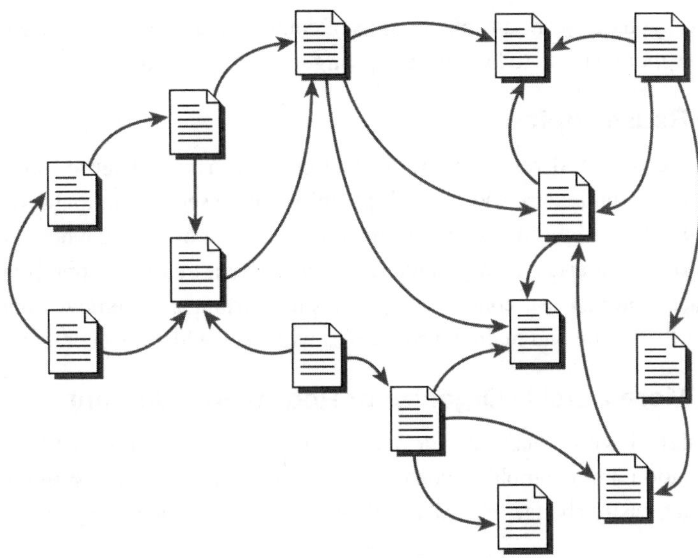

Figure 1.1 Linked and organized topics that form a web of information.

Advantages of Writing in Topics for Writing Teams

Creating technical information in topics not only helps your users find and use your information more easily, but topic-based information also has significant benefits for writers, information architects, and editors.

Writers Can Work More Productively

In your organization, several writers might work on content for related features or functional areas. To be more productive, each writer can contribute specific topics that support those related features or functions.

For example, for a complex installation guide for an enterprise database system, each writer can own some of the installation information. One writer might own information about setting up security, and another writer might own information about planning for the installation.

When you write in topics, it's easier and more productive for multiple writers to contribute to the larger set of information.

Writers Can Share Content with Other Writers

The larger a file is, the more difficult it is for you and other writers to work on content in the same file. Imagine a file that is 50 pages. Every time you work on that file, 50 pages of content are locked and unavailable for other writers to work on.

With topics, you can work on smaller units of content at any time because each topic is one file, which allows more writers to have access to larger portions of information.

Writers Can Reuse Topics

Organizations save time and resources by reusing content. You can reuse topics for multiple products, for different audiences, and for multiple information sets and output formats.

For example, you might include the same topic in both a book and a help system or share topics among product libraries. If the shared topic has content that isn't appropriate or required for every product, you can use conditional processing attributes to remove content that's not applicable rather than write and maintain two topics with nearly identical content.

Writers Can More Quickly Organize or Reorganize Content

Information designed with a narrative flow or book structure doesn't enable you to quickly rearrange information. For example, if you create separate task topics for assembling a motorcycle engine, you can easily change the order of those task topics if the motorcycle engine's design changes.

Reviewers Can Review Small Groups of Topics Instead of Long Books

Instead of asking editors, information architects, or technical experts to review long books, you can submit small sets of topics or even single topics to reviewers throughout the development cycle of the product. You're more likely to get better feedback if the reviewer can read through a handful of shorter topics than have to comb through a 100-page chapter.

DITA Topic Types

Users of technical information often need at least three types of content: procedures, background or conceptual information, and quick reference information. What most users don't need is a jumbled mix of those three types of information—for example, procedures buried in a long section of conceptual information or a table of miscellaneous commands thrown in the middle of a procedure.

Figure 1.2 shows how mixing information types can make specific pieces of information difficult to find. The topic uses a title that seems to indicate that the information is conceptual, and the topic does describe how nuclear energy is created. However, the topic then digresses into a task that describes how to connect an espresso machine to a nuclear reactor. And to make matters worse, the topic also contains a table of commands, none of which are related to how energy is created from nuclear fusion. Most users do not expect to find task information or even a list of commands in a topic called "Nuclear fusion as a power source."

Nuclear fusion as a power source

Fusion power is the process by which multiple atomic particles join together to form a heavier nucleus. This is the power source for the Exprezzoh 9000N.

This process creates the release or absorption of energy. Iron and nickel nuclei have the largest binding energies per nucleon of all nuclei and therefore are the most stable. The fusion of two nuclei lighter than iron or nickel generally releases energy while the fusion of nuclei heavier than iron or nickel absorbs energy.

The opposite process is called nuclear fission. However, because the Exprezzoh 9000N requires so much energy, it uses the nuclear fusion process.

You can connect the Exprezzoh 9000N to any operating nuclear fusion plant in a 200 mile radius of your home. To connect to a fusion reactor:

1. Obtain the required permits from your local nuclear federal agency.
2. Set up the monitoring system that is required by the federal agency.
3. Ensure that the nuclear power source light is green.

   The power source display panel is near your circuit-breaker panel for your house. The nuclear power display panel was installed when you installed the nuclear reactor piping cables.
4. Plug in the Exprezzoh 9000N to the fusion reactor power cord.
5. Start the Exprezzoh 9000N by turning the ON/OFF switch to the ON position.

After you connect the nuclear reactor, you can monitor power flow from your personal computer. You can also redirect some of that power to other appliances.

Task	Command	Example
Get reports at various intervals to view power consumption.	`exprezzoh report powerLevel` *`time`* For *time* ,you can specify to get reports every hour, every 12 hours, every day, every week, or every month: **hour** A report is created every hour. **12 hour** A report is created every 12 hours.	`exprezzoh report powerLevel 12 hour`

Figure 1.2 A topic with conceptual, task, and reference information.

To make it easier to create and deliver information that effectively separates content by type and purpose, DITA provides three main topic types: concept, task, and reference:

- A task topic describes one procedure.
- A concept topic defines what something is or how a process works.
- A reference topic contains one type of reference information that users might need as they perform tasks.

Table 1.1 shows the differences between concept, task, and reference topic types by their titles.

By separating content by type, you prevent users from wading through information that they don't need. For example, when you want to install a home entertainment system for the first time, you don't need to know about all the buttons on the remote control (reference information). Rather, you need the installation instructions. By separating the reference information from the task information, users can more quickly install their system.

Table 1.1 Examples of Topic Types by Title

Concept Topic Title	Task Topic Title	Reference Topic Title
User roles	Create user roles	Supported types of roles
High-definition television	Install a high-definition television set	Television set accessories
Espresso	Make espresso drinks	Espresso drink ingredients
Cat behavior	Herd cats	Domestic cat varieties
Databases	Configure databases for enterprise systems	Database types
Photography	Take photographs of landscapes	Digital camera models and compatibility matrix

Separating content by type also helps specific users find the information they need. For example, novice users are more likely to need conceptual information, whereas experts probably go straight to the procedures and reference material and might not bother with the concepts.

Task Orientation

Separating your content by type isn't always sufficient to make your information address the needs of your users. To help users accomplish real-world goals by using your products, you must create task-oriented topics. For example, the users of your products might want to accomplish the following real-world goals:

- Processing loan applications for a bank
- Reducing power costs by installing solar panels on rooftops
- Increasing revenues by making business processes more efficient
- Making electronic medical records available to physicians
- Manufacturing audio components for cars
- Setting up enterprise email systems so that employees can be more productive

The job of the technical writer, editor, and information architect is to create information for products that help users accomplish these goals. The goal is not simply to describe how the product works.

To create effective task-oriented information, follow these guidelines:

- Focus on the goals of the user, not the way the product works.

- Write from the user's point of view and write in active voice.
- Target the appropriate audience.
- Tell users why they need or should perform the task.
- Break up large or complex tasks into shorter subtasks and organize task topics in logical order.
- Don't bury a task in conceptual or reference information.

By applying the principles of task-oriented writing to your information, you can write your topics according to what tasks users perform rather than by the way the product works or by the way it's designed.

For more information about task-oriented writing, see "Task Orientation" in *Developing Quality Technical Information* by Hargis et al.

Task Analysis

When human factors engineers design new products, they do a task analysis so that they can understand the goals of their users and how those users want to use the product. Technical writers can also do a task analysis to understand how users work with the product. A task analysis can help you create effective task-oriented information.

During a task analysis, you need to find as much information as you can about how users currently work or intend to work with your product. A thorough task analysis can provide the following information:

- What task topics to write
- How much supporting reference and conceptual information to provide

Do a task analysis at the beginning of the project. For example, you might do a task analysis when you create information for a new product, when you reorganize a set of information, or when you model the information for a new feature, service, or technology.

TIP Project managers can also use your task analysis to scope the documentation project. Your analysis can help managers understand the size and resources required for the project.

Table 1.2 shows a task analysis of how to make an espresso drink:

The result of the analysis is that you have the following information:

- An understanding of what tasks users must complete to achieve a goal.
- An outline of what steps it takes to perform a task. Use this outline to create your DITA topics.

Table 1.2 Task Analysis of How to Make Espresso Drinks

Question	Details	Comments
What is the goal?	To make an espresso drink.	State the ultimate goal of the user; don't describe how the product works.
What tasks does the user need to perform to accomplish the goal?	• Prepare the beans. • Load the filter. • Configure the espresso machine. • Add water. • Prepare the milk. • Turn on the espresso machine.	Don't worry about order yet. Just brainstorm all possible tasks.
What are the mental and physical steps involved in each task?	Mental: Decide what kind of coffee to make. Physical: Grind beans, steam milk, and load filter.	Most tasks require mental and physical steps.
Who performs the task?	Audience: Coffee drinkers who like strong coffee drinks	Describe users in as much detail as possible.
	Experience: Advanced	
When and under what conditions is the task performed?	Requirements: Espresso machine must be configured and running; espresso drink ingredients must be available. Limitations: User must know how to create espresso drinks. Environment: Users usually make coffee in the morning and are probably sleepy.	Describe prerequisites, limitations, or restrictions to do the task.
What are potential distractions to accomplishing the goal?	Troubleshooting: Broken power source or bean grinding problem. Alternative path: Deciding to make brewed coffee rather than an espresso drink. Exception path: Missing ingredients.	Consider errors or problems that users might encounter. An exception path describes the situation when something goes wrong in the process that prevents users from completing of the task.

Table 1.2 Task Analysis of How to Make Espresso Drinks

<table>
<tr><th>Question</th><th>Details</th><th>Comments</th></tr>
<tr><td>What does the user need to know about the task?</td><td>Duration: 3 minutes if the coffee beans are ground; 8 minutes if the coffee beans are not ground.

Complexity: Easy for advanced users; medium to difficult for users making espresso drinks for the first or second time.

Frequency: Daily.</td><td>This information might impact the type of content that you include in the procedure or how you structure the task topic.</td></tr>
<tr><td>What is the sequence of tasks or steps?</td><td>Prerequisites: Install and configure espresso machine.
1. Turn on the espresso machine.
2. Prepare the beans:
a. Select the coffee beans.
b. Grind the beans to fine or extra fine.
3. Load the ingredients:
a. Load the ground beans.
b. Add the water.
4. Allow the ground coffee to brew.
5. Pour the espresso into a cup.
6. Optional: Add sugar or other ingredients.</td><td>Organize the tasks or steps in the proper order.</td></tr>
<tr><td>What is the expected result?</td><td>Make a perfectly crafted espresso drink.</td><td>State the results that users will expect to see or accomplish.</td></tr>
</table>

Although a task analysis might seem time consuming, the effort can likely save you time over the entire release cycle of your product. Use the task analysis form that appears at the end of this chapter to do an analysis for your product, service, or technology.

You can use professional modeling tools to do a task analysis, such as UML applications or the DITA-aware IBM Information Architecture Workbench. You can also use a spreadsheet or pen and paper to track the analysis questions and responses.

Minimalist Writing

No discussion of topic-based writing would be complete without a few words...very few...about minimalist writing. By following minimalist writing principles, you can create more effective topics. In minimalist writing, you should provide only the information that users need, when they need it, and nothing more.

You can find many excellent books and articles that describe minimalist writing in more detail, but remember these important principles: know your audience, remove nonessential content, and focus on user goals.

Know Your Audience

You must understand your users' level of expertise with the product, service, or technology. Do a task analysis so that you know exactly what information users need to accomplish real goals.

Analyze customer support feedback and conduct usability tests to learn about your users:

- What are users likely to know about your product or technology? For example, if your product describes how to create web pages, do you need to explain the basics of browsers?
- What goals do they want to accomplish?
- Will they understand the terminology used for your product? For example, will most of your users who are experienced with search engines know what you mean by "Boolean operators"?
- How much help will they need to resolve problems with the product? How much troubleshooting information should you provide?

Remove Nonessential Content

As a technical writer, you become an expert in the product, service, or technology that you document. Although you might understand your product well, consider what information is essential for your users. Don't create information that users don't need or care about.

This advice might sound like a no-brainer, but you often read technical information that describes the toolbar icons of a simple software product or a step in a task that says, "Type your name in the name field." If your product requires explanation of the toolbar or how to enter a name in a name field, consider improving the product interface rather than creating topics that will probably never be read.

Focus on User Goals, Not Product Functions

Avoid writing task topics that are solely about how the product works. For example, instead of writing a task topic called "Using the User Profile dialog box," which focuses on how the product works, create a task called "Changing user profiles," which focuses on a real goal.

Even for complex enterprise products, such as database or search engine systems, avoid writing lengthy discussions about how the product works. Your users want only enough information to help them set up and use their systems or products.

For example, instead of writing long chapters about how authentication works, start with tasks that show users how to set up security, and explain the options and the benefits of setting up security for various scenarios as they progress through the tasks. You can provide some concept topics, but provide just enough to get users started. Very few readers will be patient enough to read 25 or 50 pages of conceptual material before they start a task.

For more information about writing consistent, minimalist content, see *The IBM Style Guide: Conventions for Writers and Editors* by DeRespinis, et al.

To Wrap Up

DITA provides a flexible, yet rigid, framework that helps you create effective technical information. DITA is flexible in that it helps you reuse, reorganize, and create content quickly. It's rigid in that you must be disciplined enough to adhere to the principles of good topic-based writing to take advantage of the benefits of DITA.

Writing task-oriented DITA topics takes some practice, but when you do create those perfect topics, your users can:

- Find the information they need faster
- Accomplish their goals more efficiently
- Read only the information they need to read

Writing in DITA not only benefits your users, but it also benefits you as a writer. By using DITA to create topic-based information, you can:

- Maintain and reuse topics more effectively
- Organize or reorganize topics more quickly
- Share and distribute the work on topic files more easily, which increases writer productivity

Topic-based information also starts with a thorough knowledge of your users and their goals. Do a comprehensive task analysis before you write anything. Don't be afraid to spend more time on planning your information and analyzing the tasks that users will want to do with your product.

By writing your content in discrete, task-oriented DITA topics, you can create useful, coherent, and retrievable technical information.

Topic-Based Writing Checklist

Guideline	Description
Check that each topic contains only one type of information.	Use the different topic types in DITA to separate task, conceptual, and reference information.
Check that each topic is self-contained.	Eliminate text that points to other topics (other than the occasional cross-reference), text that points to sections in the topic, such as "The next section describes table spaces," or topics that serve merely to glue other topics together.
Check that topics don't cover too much information.	For example, writing one topic to describe even a single commercial aircraft model is too much information for one topic. Instead, break up the topic into more concept topics and subtopics, such as "Engines," "Passenger seating," "Pilot console," and "Electrical systems."
Ensure that your content is task oriented.	Task orientation means that your content focuses on real user goals, not product functions or features. Don't focus on pseudo-tasks such as "Using the email feature." Using something isn't a real-world goal. "Sending a document through email" is a real goal.
Do a task analysis to decide what information your topics should contain.	Technical information is effective only if it answers users' questions quickly and completely. Ensure that you understand the level of expertise of your users and their goals.
Follow minimalist guidelines.	• Know your audience. • Eliminate nonessential content. Write only the content that users need. Don't paper the product. • Focus on user goals, not product functions.

Task Analysis Form

Use this form to do a task analysis on a new product, feature, or component.

Question	Details	Comments
What is the business goal?		
What tasks does the user need to perform to accomplish the goal?		
What are the mental and physical steps required for each task?	Mental: Physical:	
Who performs the task?	Audience: Experience: Role: Authority:	
When and under what conditions is the task performed?	Requirements: Limitations: Environment:	

Question	Details	Comments
What are potential distractions to accomplishing the goal?	Troubleshooting: Alternative path: Exception path:	
What does the user need to know about the task?	Duration: Complexity: Frequency:	
What is the sequence of tasks?		
What is the expected result?		

CHAPTER 2

Task Topics

Task topics are the true worker bees of technical information. They're what distinguish technical writing that is designed for products, technologies, and services from mystery novels or essays on nature.

The DITA task topic provides a wonderfully useful structure to help you create effective procedural information. Task topics include many unique XML elements that you won't see in concept or reference topics. These elements provide the structure for content such as prerequisites, steps, examples, and other items that you typically create for procedures.

Task topics often need supporting concept and reference topics, but users typically run straight to the task topics and read conceptual and reference material only when they need to. For example, in a users' guide that describes how to train dogs, you might create a task topic such as "Teach simple commands." To support that task, you might create a concept topic called "Domestic dog behavior" and a reference topic called "Dog breeds and amenability to training." Together, these topics help users complete their goal: training their canines to do as they're told.

Writing effective task topics is critical if you want to provide task-oriented information that helps users accomplish real-world goals.

When you write task topics, follow these general guidelines:

- Separate task information from conceptual or reference information.
- Write one task per topic.
- Create tasks and subtasks to organize long procedures.

Separate Task Information from Conceptual or Reference Information

Separate tasks from long conceptual or reference information so that tasks are short, retrievable, and reusable. If you overload a task topic with too much conceptual or reference information, expert users get frustrated because they must wade through information that they might already understand.

Also, ensure that a task isn't buried in paragraphs of conceptual or reference information. Users won't expect to find a task buried in a table of reference material. And they'll probably be upset that they spent so much time digging through information that they didn't need in order to find that one specific task.

Write One Procedure per Topic

Write only one procedure per task topic so that you can more easily manage, organize, and reuse these topics and so that your users can find specific tasks when they need them. For example, don't combine the tasks for installing and uninstalling software in a single topic. Users typically won't need both of these procedures at the same time.

Also, adding more than one task per topic makes those secondary procedures harder to find. When you transform the DITA topics to an output format such as HTML, you'll see only the title of the first procedure in the table of contents.

Create Subtasks to Organize Long Procedures

Procedures that are more than 10 steps can be difficult to follow, especially if some of the steps are complex or have many substeps. Instead of creating one long procedure in a single topic, break it up into several shorter task topics. You can then assemble those task topics into a logical order that helps users to finish the entire procedure.

To create a set of task topics, start by creating a parent topic, or supertask, that describes the overall task flow. Then, nest the child task topics under the supertask in a logical order. The output will show users what tasks they need to follow and in what order those tasks must be completed.

For example, to describe how to install a database system for a financial services business, you'd need a dozen or more task topics. With DITA, you can organize and link those task topics so that users see a clear sequence of tasks. You wouldn't want to create a single task topic with 100 steps. Think of your poor users!

The following example divides the task of setting up a nuclear fusion power source for an espresso machine into three separate task topics:

1. Gather permits
2. Identify a nuclear reactor to connect to
3. Prepare your home for nuclear power

To organize these tasks, you can create a supertask topic called "Prepare to install Exprezzoh 9000N" and nest the subtask topics under that topic, as in Figure 2.1. When the DITA files are transformed to HTML, you get the following output.

Prepare to install the Exprezzoh 9000N

Before you install the Exprezzoh 9000N, ensure that you have the necessary permits to connect to a nuclear facility and that your home is prepared for nuclear power.

1. Gather and mail mandatory permits
   Before you can connect to a nuclear reactor, you must have several permits from federal, state, and local governments.
2. Identify a nuclear reactor to connect to
   After you obtain the necessary permits to connect to a fusion reactor, you must identify a reactor that is within 100 miles of your home.
3. Prepare your home for connecting to a nuclear reactor
   Ensure that your family, pets, neighbors, and all living things within 100 mile radius of your home are protected before you connect the espresso machine to nuclear power.

Parent topic: Install, configure, and make beverages with the Exprezzoh 9000N

Figure 2.1 The HTML output of a sequence of task topics.

By dividing long procedures this way, your users aren't overwhelmed, and they can see the overall sequence of tasks. You can also more easily reuse these shorter task topics elsewhere in your information set if necessary.

For more information about DITA maps and organizing parent and child topics, see Chapter 6, "DITA Maps and Navigation."

Task Components and DITA Elements

Unlike concept or reference topics, which use <section>, <p>, <table>, and other common elements, task topics include unique elements such as <steps>, <cmd>, <choicetable>, and <prereq> elements to help you structure procedural information.

The challenge is to understand what content to add to these task topic elements and to use these elements consistently. Table 2.1 describes common elements used in task topics.

Table 2.1 Common DITA XML Elements in DITA Task Topics

DITA Element	Description
<title>	Provides the topic or section title
<shortdesc>	Introduces the task
<steps>	Contains ordered steps for the entire procedure
<steps-unordered>	Contains unordered steps for the entire procedure
<context>	Provides background information that users need to understand to do the task
<steps>, <step>, and <cmd>	Contains the steps for a task
<substeps>, <substep>, and <cmd>	Contains the substeps for a step
<info>	Provides additional information that users might need when they complete a step
<stepresult>	Provides what happens when a step is completed
<stepxmp>	Provides an example of what happens when a step is completed
<choices> and <choice>	Displays choices in a bulleted list instead of ordered steps
<choicetable>	Displays choices in a two-column table where you can describe steps for each choice
<postreq>	Provides information about tasks that must be done after the current task is done
<example>	Provides an example that illustrates or supports the task
<result>	Provides the expected outcome when the task is done

Titling the Task: <title>

Distinguish task topics from concept and reference topics by using a gerund, imperative verb phrase, or a "how to" phrase in the task title.

Rather than the title "Exprezzoh 9000N installation" for a task topic, use a title such as:

- "Installing the Exprezzoh 9000N"
- "Install the Exprezzoh 9000N"
- "How to install the Exprezzoh 9000N"

Users need to distinguish task topic titles from concept or reference topic titles. For example, you might have a concept topic titled "Databases" and its related task titled "Configuring databases."

If you're consistent with the way you title task topics and you use a verb-based or "how-to" title, users will know what type of information to expect just by looking at the title.

For more information about titles for topics, see the "Task Topics" in *The IBM Style Guide* by De Respinis et al.

Introducing the Task: <shortdesc>

Introduce the task in the <shortdesc> element, but keep short descriptions short. If you need to add more background information about the procedure, add it to the <context> element.

In the <shortdesc> element, you can introduce tasks by describing:

- An overview of the procedure
- The benefits or importance of the task
- Limitations or requirements
- Brief conceptual information

For more information about writing effective short descriptions, see Chapter 5, "Short Descriptions."

Adding More Background Information: <context>

The main component of a task topic is the task body, or <taskbody> element. Nested in the <taskbody> element is a <context> element where you can add more background information to complete than what's in the <shortdesc> element.

Orient your users by providing just enough context such as where and when to perform the task or to explain the benefit of performing the task. However, don't delve into long, complex concepts. For long concepts, create a separate concept topic and link it to your task topic.

If the task is short or relatively simple and you provide an effective introduction in the <shortdesc> element, omit the <context> element. You also might include a step introduction in the <context> or <stepsection> element to make it easier for users to quickly locate the procedure.

TIP In DITA 1.2, use the <stepsection> element to include an introduction to the steps of a task. The <stepsection> element is not available in DITA 1.1.

Figure 2.2 shows a task topic that includes a step introduction for a topic called "Cleaning the fuel injection system."

If you use step introductions, be sure you write and apply them consistently across your information set. For example, don't use sentence fragments, such as "To clean the fuel injection system:" in some tasks and full sentences, such as "To clean the fuel injection system, complete these steps:" in others.

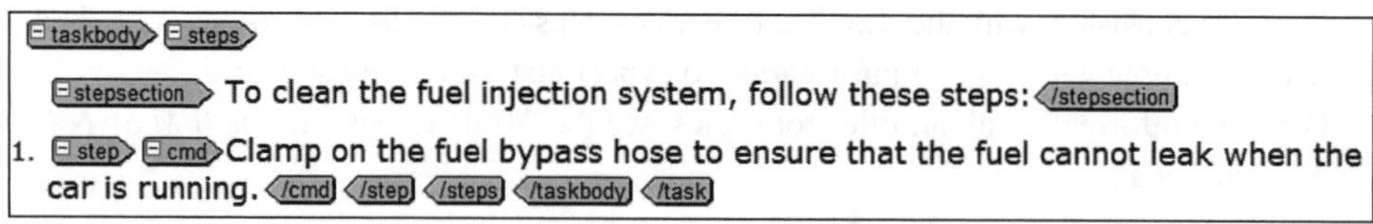

Figure 2.2 A <stepsection> element that's used to introduce a task.

Alternatively, you can omit the introduction to the steps and modify your XSL transform to include a section heading, or label, that separates the text in the <context> element from the steps. For example, use the label "Procedure" to help users quickly find the steps.

Figure 2.3 shows the HTML output of a topic with the label "Procedure" before the steps of a task and doesn't include an introductory sentence before the steps.

Gather and mail mandatory permits

You can use the permit forms that are provided with the espresso machine. Before you can connect to a nuclear reactor, you must have several permits from federal, state, and local governments.

Procedure

1. Print and complete the following forms:
   - NUK1234F to obtain permission from the federal agency to use the reactor.
   - NUK4321S to obtain permission from the state agency to use the reactor.
   - NUK2244L to obtain permission from the county agency to use the reactor.
2. Mail the form to the address listed on the top of the form. You should receive approval in about 6-24 weeks.

The returned approvals contain codes to enable the nuclear reactor.

You will be prompted to enter these codes. Enter the codes that are on the returned approval forms when you enable the espresso machine to use nuclear power for the first time.

Figure 2.3 A task with the label "Procedure" before the steps.

Describing Prerequisites: <prereq>

Be sure to describe what users need to know and what conditions need to be met before they begin a task. Describe these dependencies in the <prereq> element. You can use many common elements in the <prereq> element, such as ordered lists, unordered lists, and paragraphs.

Be careful not to add a long procedure in the <prereq> element. Instead, create another task topic for the prerequisite procedure.

In Figure 2.4, the topic contains effective prerequisite information for a task.

In the output, as shown in Figure 2.5, the prerequisite text is displayed as a normal paragraph and unordered list.

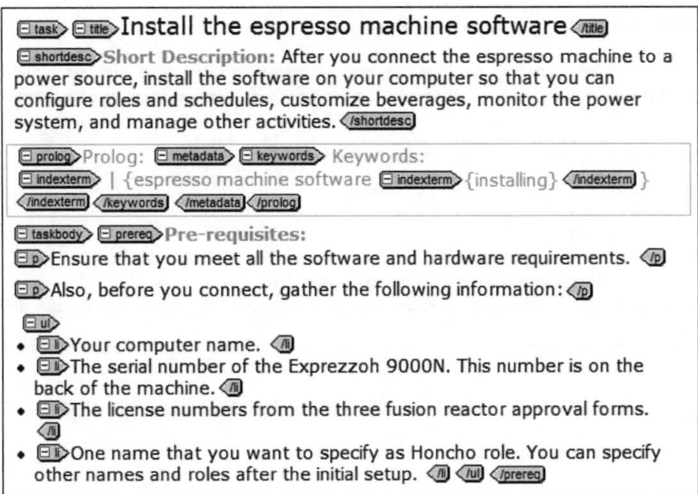

Figure 2.4 Prerequisite information in the <prereq> element.

Install the espresso machine software

After you connect the espresso machine to a power source, install the software on your computer so that you can configure roles and schedules, customize beverages, monitor the power system, and manage other activities.

Ensure that you meet all the software and hardware requirements.

Also, before you connect, gather the following information:

- Your computer name.
- The serial number of the Exprezzoh 9000N. This number is on the back of the machine.
- The license numbers from the three fusion reactor approval forms.
- One name that you want to specify as Honcho role. You can specify other names and roles after the initial setup.

Figure 2.5 Prerequisite information in HTML output that's shown as a paragraph and list before the steps in a procedure.

BEST PRACTICE Write prerequisites as imperative statements that call users to action. Include links to tasks that users should complete before beginning the current task.

You can modify your XSL transform to include a label of your choice to identify prerequisite information in the output. For example, in Figure 2.6, the topic in HTML shows the label "Before you begin" before the steps.

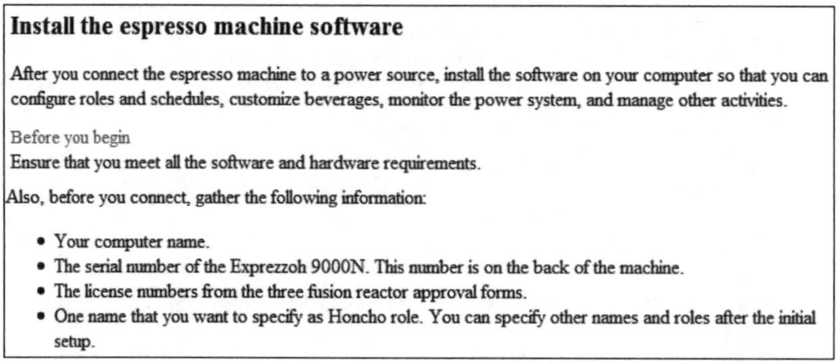

Install the espresso machine software

After you connect the espresso machine to a power source, install the software on your computer so that you can configure roles and schedules, customize beverages, monitor the power system, and manage other activities.

Before you begin

Ensure that you meet all the software and hardware requirements.

Also, before you connect, gather the following information:

- Your computer name.
- The serial number of the Exprezzoh 9000N. This number is on the back of the machine.
- The license numbers from the three fusion reactor approval forms.
- One name that you want to specify as Honcho role. You can specify other names and roles after the initial setup.

Figure 2.6 A prerequisites section that includes an automatically generated label of “Before you begin.”

Writing the Procedure: <steps> and <steps-unordered>

In DITA, you can describe two types of tasks in a task topic: procedures with ordered steps or procedures with unordered steps.

To set up ordered steps, use the <steps> element, which is the outermost container for your steps. Use this element for most of your tasks, as shown in Figure 2.7.

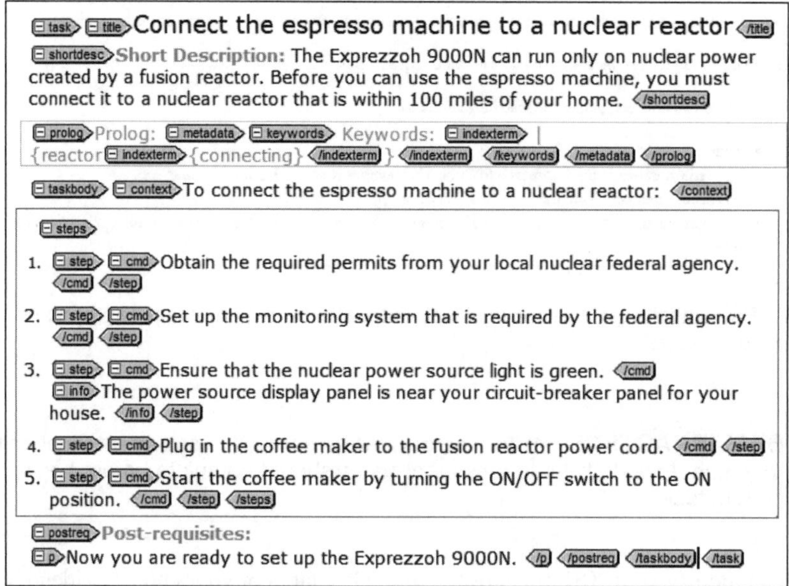

Figure 2.7 A task that uses the <steps> element.

To set up unordered steps, use the <steps-unordered> element. The default output shows a bulleted list instead of numbered steps. You probably won't write many tasks that include only unordered steps, such as the following task in Figure 2.8.

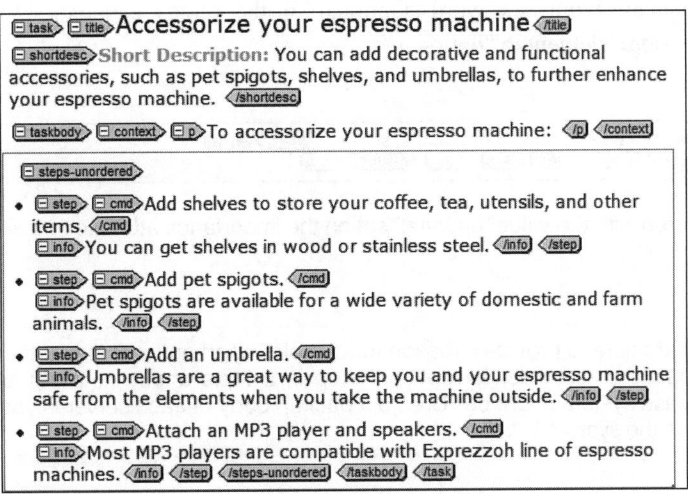

Figure 2.8 A task that uses the <steps-unordered> element.

Whether you create ordered or unordered tasks, you write each step in the <step> and <cmd> elements. If needed, you can further clarify each step by adding information in other elements, such as <info>, <choicetable>, <stepresult>, <stepxmp>, and <substeps> elements.

Writing the Steps: <step> and <cmd>

When you write steps in a task topic, follow these guidelines:

- Include an imperative sentence in every <cmd> element, such as "3. Pour the coffee beans into the grinder." Ensure that a step never contains only an indicative statement such as "3. Now you pour the coffee beans into the grinder."
- Write one step per significant action and combine short, simple steps. For example, combine "4. Click OK" and "5. Click Save" into one step: "4. Click OK and click Save."
- Even if your task has only one step, add the imperative sentence in a <cmd> element as you would if you were writing a task with several steps. When you transform the DITA source file to an output format such as HTML or PDF, a number won't be added for a task with a single step.

Steps can also be conditional or optional. A *conditional step* is a step that's required only if some condition is true; for example, "If the espresso machine is not already started, press the

Power button." Use a verbal cue such as an "if" statement to indicate that a step is conditional. The <step> element has no attributes that you can use to indicate that a step is conditional.

An *optional* step is one that doesn't need to be completed for the task to be successful. Figure 2.9 shows how you can use the importance attribute to specify that a step is optional. Rather than manually typing the word "Optional" in the <cmd> element, set the importance attribute of the <step> element to "optional."

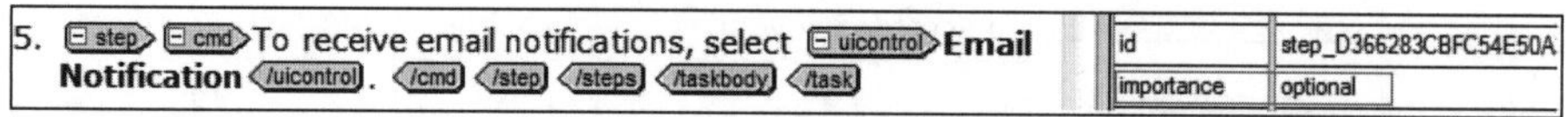

Figure 2.9 A step with the value "optional" set on the importance attribute of the <step> element.

TIP Be sure to provide a reason for completing an optional step if the reason isn't obvious. For example, an optional step in a topic about installing an enterprise database system might be "Create a backup copy of each server in your enterprise in case the system fails."

In Figure 2.10, you see the automatically generated label "Optional:" before the step in the output.

5. **Optional:** To receive email notifications, select **Email Notification**.

Figure 2.10 A step with the label "Optional."

Describing Choices Instead of Steps: <choices> and <choicetable>

If you want to include choices for a specific step in your procedure, use the <choices> or <choicetable> element, which you can insert in any <step> element. However, you can't insert them in a <substep> element.

A common question is "What's the difference between the <choices> element and the <choicetable> element other than one is a bulleted list and the other is a table?" Use the <choices> and <choicetable> elements in the following situations:

- Use the <choices> element for simple one-part items such as a list of options that users select. The items in a <choices> element will be rendered in output as a bulleted list.
- Use the <choicetable> element for two-part items such as the name of an operating system and its command or a motorcycle model and its oil filter part number. The items in a <choicetable> element will be rendered in the output as—you guessed it—a table.

Figure 2.11 contains choices to indicate that the user must select one or more items in the list.

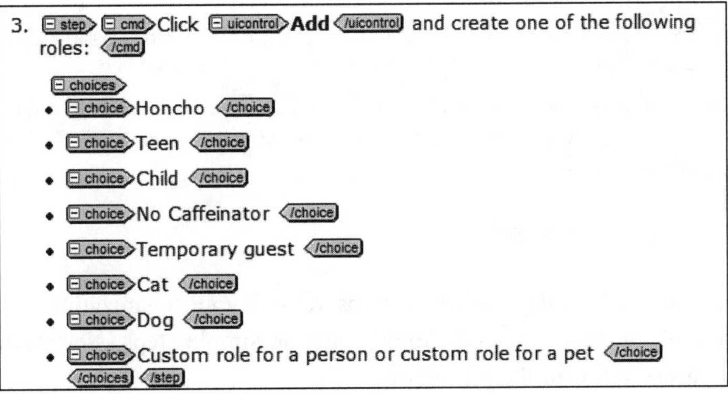

Figure 2.11 List of options in the <choices> element.

Figure 2.12 uses a <choicetable> element because the command has two parts: the operating system followed by a specific command for that operating system.

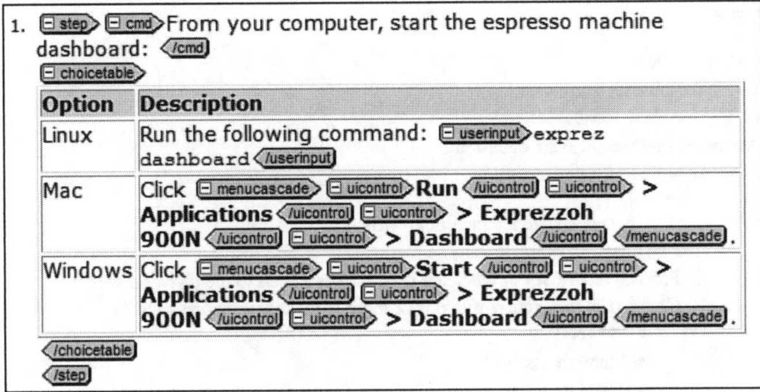

Figure 2.12 A list of operating systems and their corresponding commands in a <choicetable> element.

Including Substeps: <substeps>

You can add up to one level of ordered substeps to any step. By default, substeps are rendered in the output as lettered steps. You insert the <substep> element after the <cmd> element of the main step.

Figure 2.13 shows substeps under step 4.

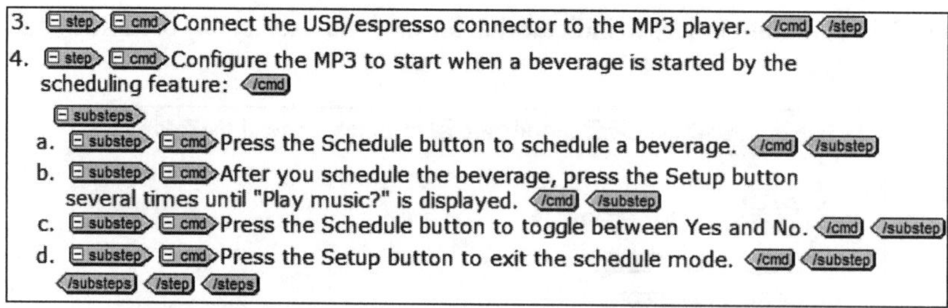

3. <step> <cmd>Connect the USB/espresso connector to the MP3 player. </cmd> </step>
4. <step> <cmd>Configure the MP3 to start when a beverage is started by the scheduling feature: </cmd>
<substeps>
a. <substep> <cmd>Press the Schedule button to schedule a beverage. </cmd> </substep>
b. <substep> <cmd>After you schedule the beverage, press the Setup button several times until "Play music?" is displayed. </cmd> </substep>
c. <substep> <cmd>Press the Schedule button to toggle between Yes and No. </cmd> </substep>
d. <substep> <cmd>Press the Setup button to exit the schedule mode. </cmd> </substep>
</substeps> </step> </steps>

Figure 2.13 Substeps under step 4.

If you need substeps inside of other substeps, rethink your organization: You should break the task into multiple subtasks, or you should combine simple steps. Nesting too many steps makes it hard for users to follow the procedure.

For example, the task in Figure 2.14, which shows only the first part of the task, is difficult to understand because it has too many levels of nested substeps.

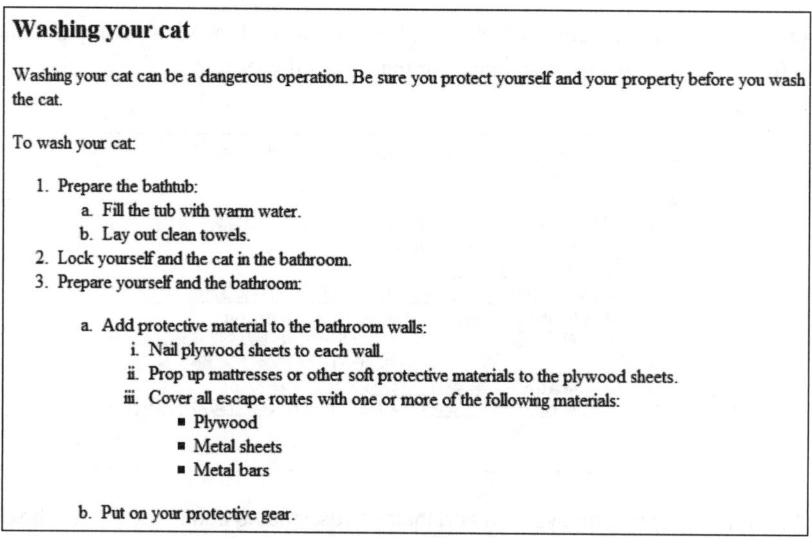

Washing your cat

Washing your cat can be a dangerous operation. Be sure you protect yourself and your property before you wash the cat.

To wash your cat:

1. Prepare the bathtub:
   a. Fill the tub with warm water.
   b. Lay out clean towels.
2. Lock yourself and the cat in the bathroom.
3. Prepare yourself and the bathroom:
   a. Add protective material to the bathroom walls:
      i. Nail plywood sheets to each wall.
      ii. Prop up mattresses or other soft protective materials to the plywood sheets.
      iii. Cover all escape routes with one or more of the following materials:
         - Plywood
         - Metal sheets
         - Metal bars
   b. Put on your protective gear.

Figure 2.14 A task that's difficult to follow because of too many levels of nested steps.

You can combine some steps and delete unnecessary steps to make the task easier to follow, as shown in Figure 2.15.

You could also break the task into two shorter tasks: "Preparing your washing area" and "Washing your cat."

Washing your cat

Washing your cat can be a dangerous operation. Be sure you protect yourself and your property before you wash the cat.

To wash your cat:

1. Fill the tub with warm water and lay out clean towels.
2. Lock yourself and the cat in the bathroom.
3. Add protective material to the bathroom walls:
    a. Nail plywood sheets to each wall.
    b. Prop up mattresses or other soft protective materials to the plywood sheets.
    c. Cover all escape routes with plywood, metal sheets, or metal bars.
4. Put on your protective gear.

Figure 2.15 A task that has only one level of substeps.

BEST PRACTICE Ensure that steps that introduce substeps have a clear imperative that states the goal of the substeps, such as "Add protective material to the bathroom wall." Avoid using sentences such as "Follow these steps" as the introduction to substeps because such sentences give users no idea what action the substeps will describe.

Describing the Outcome of a Step: <stepresult>

To describe the result of an action in a step or substep, use the <stepresult> element.

In Figure 2.16, the step includes a <stepresult> element that describes what happens when users press the Setup button.

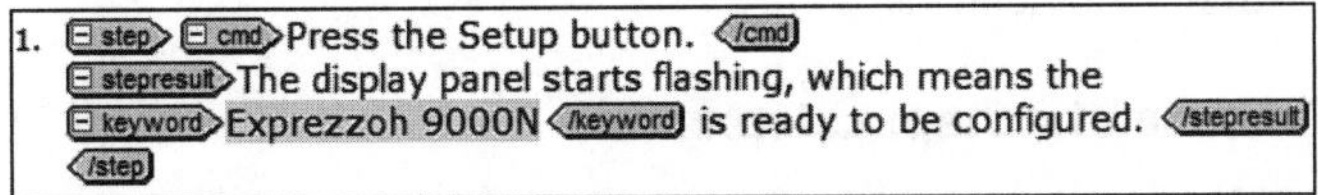

Figure 2.16 Information in the <stepresult> element that describes what happens when the step is completed.

WATCH OUT In software documentation, avoid ending every step with a sentence that says that some window or page opens. For example, if a step says "To save a new document, click Save As," you rarely need to follow that step with "The Save As dialog box opens." If the user's context has changed because of completing a step, introduce the next step by setting the context. For example, if you think users might be confused about where they are in the interface, start that next step with "In the Save As dialog box, enter a name for the file and click Save."

Including Additional Information for a Step: <info>

The <info> element can contain information that you want to add about a step or substep. Be careful not to add long conceptual information in the <info> element or add details that users won't care about.

Use the <info> element only if the information that you want to add doesn't fit logically into the <stepresult> or <stepxmp> element.

In Figure 2.17, the <info> element contains information about a power source panel.

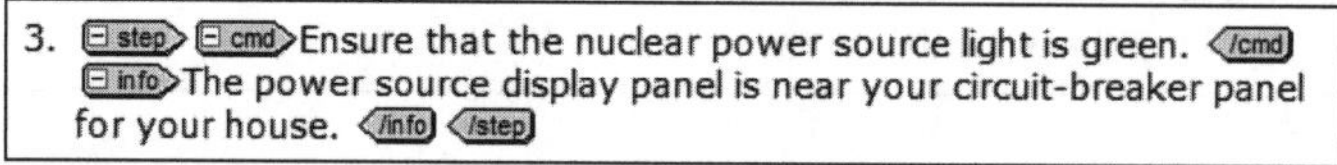

Figure 2.17 A step that includes additional information in the <info> element.

You can also insert other types of elements and text in the <info> element, as shown in Figure 2.18.

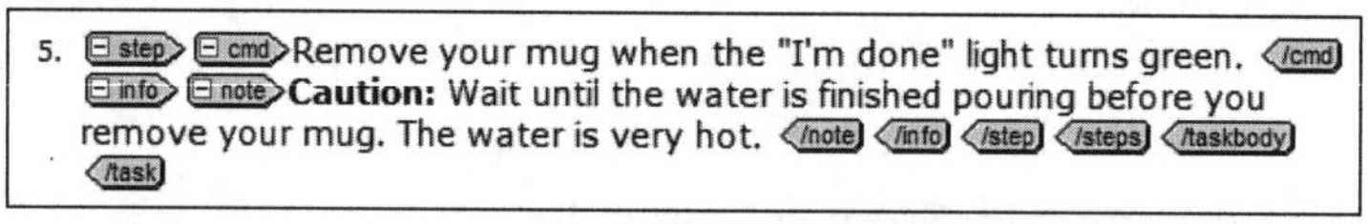

5. step cmd Remove your mug when the "I'm done" light turns green. /cmd
info note **Caution:** Wait until the water is finished pouring before you remove your mug. The water is very hot. /note /info /step /steps /taskbody /task

Figure 2.18 A step that includes a <note> element in the <info> element.

Remember to add the imperative statement in the <cmd> element only. Don't add more instructions in the <info> element.

Providing an Example of a Step: <stepxmp>

If you want to provide an example of what to do in a step or substep, use the <stepxmp> element. The example should help users understand how to complete the step.

For example, the step shown in Figure 2.19 contains a description of how users enter a time.

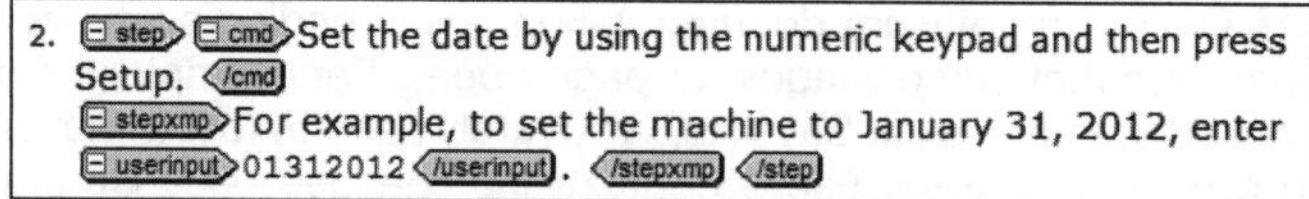

Figure 2.19 Text in the <stepxmp> element that describes how to complete the step.

Concluding the Task: <example>, <postreq>, and <result>

After the steps of the task, you can conclude the task by adding:

- An example that supports the entire procedure: <example> element
- Information about what to do next: <postreq> element
- A description of what the task helps the user accomplish: <result> element

Avoid using these elements simply because they're available. You don't need to conclude every task by simply restating the title or short description in the <example>, <postreq>, or <result> elements. For example, you rarely need a concluding statement such as "You have now configured your database system."

Describing an Example That Helps Users Complete the Task: <example>

Good examples are concrete and are sometimes more informative than the steps of a task. At the end of a task, you can add an example that illustrates or supports that task. Task topic examples might include:

- A code sample that shows how a software application is created
- A configuration of hardware that shows users how to assemble that hardware for a particular environment

You can add a title in the <title> element for the example to make the example easier to find. Use specific titles. For example, instead of writing an uninteresting title such as "Example," use a more specific title such as:

- "Sample MyWorld application"
- "Example of DITA tagging in task topics"
- "Example: Speaker configuration for your home entertainment system"

In Figure 2.20, the user completes the task of setting up roles for the Carey family. The example in the <example> element helps users complete the task of setting up roles by showing how sample roles might be defined for each family member.

If you need to provide more extensive examples that span more than one task topic, include those examples in their own reference or concept topics. Use the <example> element in a task topic only for an example that illustrates how users do one specific task.

Describing What to Do After the Task Is Completed: <postreq>

In the <postreq> (postrequisites) element, you can describe what users must do after they complete a task. For example, if the user finished setting up printer hardware, you can briefly describe or link to the task that describes how to install the printer software.

Be sure not to add a second long procedure in the <postreq> element. Instead, create a new task topic. Then, in the first task topic, briefly describe that task or link to it in the <postreq> element.

<example> <title>**Example: roles for the Carey household**</title>
<p>The Carey household has two parents, one live-in grandfather, three kids under the age of 18, two cats, one dog, and one pet rat. Mom and Dad set up the following roles for the family:
<table>
<tgroup>

Family member	Role
Mom	Honcho
Dad	Honcho
Grandpa	Custom role called "Gramps"
Francine: child, 12 years old	Child
Frank: child, 17 years old	Teen
Frieda: child, 8 years old	Child
Malluco: cat	Cat_1
Skeeter: cat	Cat_2
Jeff: dog	Dog_1
Templeton: rat	Custom role called Rat_1

</tgroup>
</table> </p> </example> </taskbody> </task>

Figure 2.20 Examples of role names in the <example> element.

Figure 2.21 uses a <postreq> element to describe what users need to do after they complete the task called "Gather and mail mandatory permit forms."

Describing the Results of Completing the Task: <result>

In the <result> element, you can describe what happens after the task is completed. For example, in a task topic called "Adjusting the idle speed on your motorcycle," you can add information about the expected results of the task in the <result> element (Figure 2.22).

In another example (Figure 2.23), the <result> element contains information about what occurs after the user completes the task of refreshing the water supply for his or her pets.

Be careful not to describe the obvious. You might be tempted to add conclusions in the <result> element of every task topic. However, the result of the task is usually clear from the title and short description. For example, if your task topic describes changing the oil on your truck, you don't need to conclude the topic by saying "You now have fresh oil in your truck."

To Wrap Up

Task topics are typically the core of technical information because users want to be shown how to use products, services, or technologies to get something done. Before you write task topics, be sure you understand your audience and their goals. Don't write task topics that focus on how the product works.

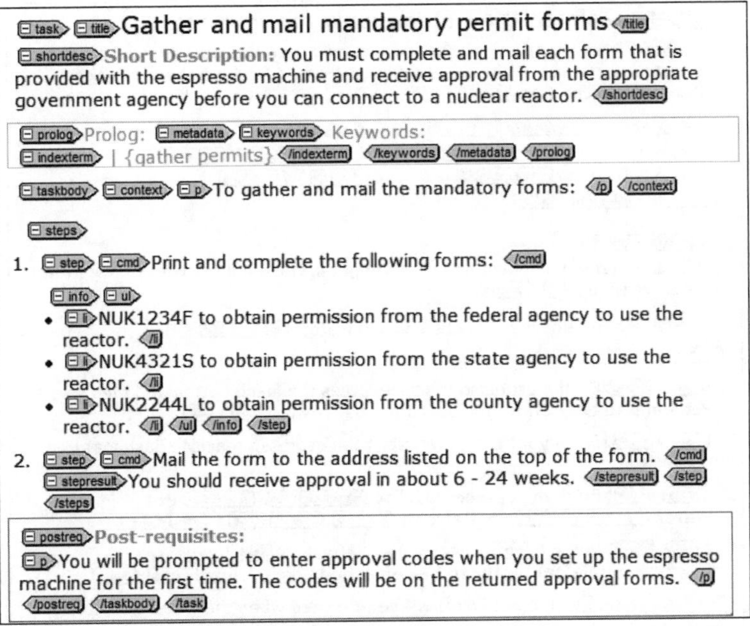

Figure 2.21 Information in a <postreq> element.

5. step cmd Turn the pilot screw half way between the low idle points to finish the adjustment. /cmd /step /steps

result Result: p After you adjust the pilot screw, the enqine should now run at 900 - 1100 RPMs. /p /result /taskbody /task

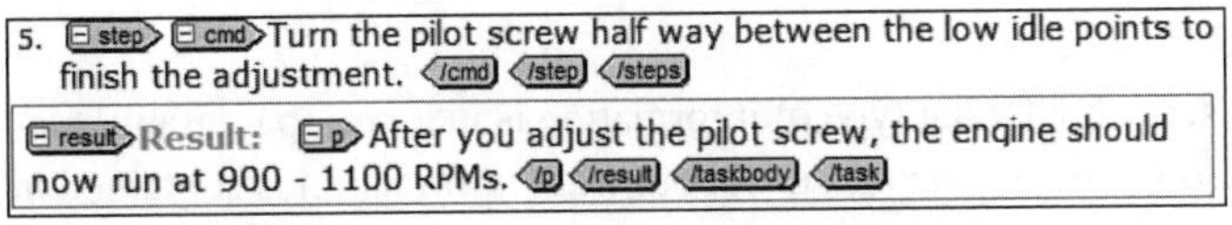

Figure 2.22 Information in a <result> element that describes the results of completing a task.

To create an effective task topic, be sure to:

- Include only one procedure per topic.
- Focus on real user goals.
- Remove long conceptual and reference information from task topics.
- Divide long procedures into shorter subtask topics.

DITA task topics also contain many unique XML elements, such as <prereq>, <steps>, <cmd>, <info>, <stepresult>, and <postreq> elements, that aren't shared by other topic types. Writing for these unique elements can be challenging when you first learn to use DITA, but these elements also provide better semantic clarity to your information. Remember that DITA is not only designed for topic-based information but that it's also a semantic markup language.

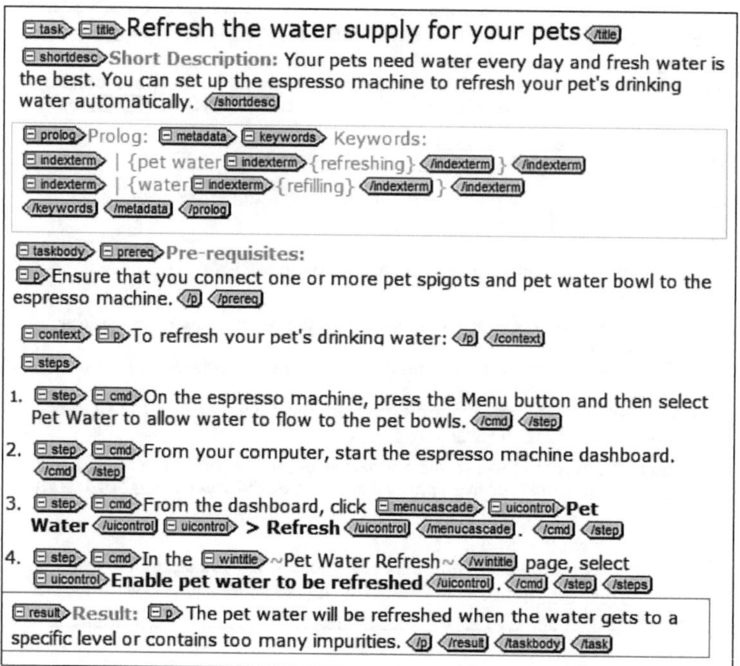

Figure 2.23 Information in a <result> element at the end of a task.

For example, you can't be sure about what type of information is in a <p> (paragraph) element other than a block of text. However, you can be sure about what information is contained in a <prereq> or <stepxmp> element.

When you write task information, be sure to write for and apply the task topic elements consistently and correctly. Your users will also appreciate getting task topics that are direct, relevant, consistent, and easy to follow.

Task Topic Checklist

Guideline	Description
Include only one procedure per task topic.	DITA allows only one <steps> element per task topic, which is meant to prevent writers from adding multiple procedures in one topic. Avoid using the <ol> element as a way to add another task.
Use verb-based or "how to" phrases for task titles.	For example, use: • "Create roles" • "Installing the home entertainment system" • "How to create espresso drinks" Whichever title style you choose, be consistent.
Create effective task topic short descriptions and introductions.	For most tasks, you can introduce them by describing: • An overview of the procedure • The benefits or importance of the task • Limitations or requirements • Brief conceptual information
Ensure that every step has an imperative verb.	In the <cmd> element, ensure that every step provides a clear action. Use an imperative sentence, for example: "1. Start the espresso machine by pressing the ON button." Be sure that steps don't contain indicative sentences as the only action, such as "1. You can start the espresso machine by pressing the ON button."
Write 10 steps or fewer per task topic.	If a task requires more than 10 steps, break the task into separate task topics and nest them in the correct order in the DITA map.
Combine short steps.	Combine short, simple steps such as "1. Click **File**" and "2. Click **New**." In this case, use the <menucascade> and <uicontrol> elements to combine the steps to create output such as "1. Click **File > New**."
Use the <steps> and <steps-unordered> correctly.	Use the <steps> element to create ordered steps for your task. Use the <steps-unordered> element if your task describes steps that can be done in any order.

<table>
<tr><th>Guideline</th><th>Description</th></tr>
<tr><td>Specify steps as optional or conditional as needed.</td><td>The differences between optional and conditional steps are:
• Optional steps don’t need to be completed. Specify the “optional” value in the importance attribute in the <step> element.
• Conditional steps must be completed in a specific situation. Because no attribute for conditional steps exists in DITA, use phrasing in the beginning of the step that indicates the step is conditional, such as “If you installed the football game software, select up to 11 players for each team.”</td></tr>
<tr><td>Don’t nest more than one level of substeps.</td><td>DITA allows only one level of substeps in a task. Nesting more than one level of substeps by adding <ol> elements in <info> or other elements makes it difficult for users to follow the task.</td></tr>
<tr><td>Use the <choice> and <choicetable> elements appropriately.</td><td>Use each element correctly:
• Use the <choices> element for simple one-part items such as a list of options, roles, part numbers, or forms that users select.
• Use the <choicetable> element for two-part items such as the name of an operating system and a corresponding file name or a bicycle model and its frame size.</td></tr>
<tr><td>Use the <info>, <stepxmp>, and <stepresult> elements correctly.</td><td>Follow these guidelines:
• Use the <info> element to describe additional information about a step.
• Use the <stepxmp> element to describe an example that shows how to complete the step.
• Use the <stepresult> element to describe what occurs or might occur after a step is done. Remember that for software documentation, you don’t need to add “The XY window opens” in the <stepresult> element in every step. Users can see what windows open when they do the step.</td></tr>
<tr><td>At the end of a task, add an example, postrequisites, and results as needed.</td><td>Follow these guidelines:
• Use the <example> element to show examples of what the procedure might accomplish.
• Use the <postreq> element to describe what users must do after the task.
• Use the <result> element to show possible outcomes of completing the task.</td></tr>
</table>

CHAPTER 3

Concept Topics

Task topics might be the core of your technical information, but without effective concepts to support those tasks, users of your product, service, or technology wouldn't get far. You can write excellent task topics for, say, how to fly small aircraft, but if the pilot doesn't understand basic concepts such as lift, drag, and thrust, let's just hope that pilot and plane never leave the hangar.

Concept topics typically explain or define some idea. They often include background information that users might need to understand before they work with a product or start a task.

Use concept topics to:

- Describe a system, product, or solution
- Outline a process
- Introduce tools and technology
- Explain features, components, characteristics, restrictions, or capabilities
- Define terms in more detail than you would in a glossary
- Describe benefits or help users to make choices between options

For example, if your users are new to enterprise email systems and must install and configure such a system, you should provide conceptual information that describes archiving, stubbing, access control lists, records management, and other relevant concepts.

TIP Write concept topics to support tasks and goals, not the other way around. Users read technical information because they want to accomplish a goal. The goal for users of most technical products is not to understand a concept but to complete a task, such as changing a tire, searching for files across multiple databases, or building medical devices.

Follow these guidelines when you write concept topics:

- Describe one concept per topic.
- Create a concept topic only if the idea can't be covered more concisely elsewhere.
- Separate task information from conceptual information.

Describe One Concept per Topic

For example, describing how modern car engines work is probably too long for one topic. You should break this information into separate concept topics, each covering subjects such as intake, compression, combustion, and exhaust. You can mention how these concepts are related in an overview topic, but separate the details of each concept into its own topic.

The advantage of separating your concepts is that users can find and read only what they need, and you can more easily reuse those concept topics elsewhere in your documentation.

When you write concept topics, be careful not to digress. You might find it tempting to add ideas that seem interesting, but digressions make your topics more cluttered and confusing. For example, don't let a description of search engine technology digress into the search habits of people who use social media.

Create a Concept Topic Only if the Idea Can't Be Covered More Concisely Elsewhere

Use a concept topic to cover an idea that requires something more than a glossary definition or a passing mention in a task topic. Be sure that you know how much your users are likely to understand before they use your product so that you don't describe what's obvious or leave out what's not obvious. You often need to include concept topics to support your task topics for novice users to get the background information that they need. Also, moving lengthy conceptual information from task to concept topics eliminates unwanted clutter from the task topic.

For example, if you write information about how to design web pages, you should include a concept topic about how web browsers find and display web pages. Providing only a cursory definition of HTML and web browsers doesn't give novice users enough information to help them build their website. Which concepts you cover and how much detail you provide depends on the expertise of your audience.

Separate Task Information from Conceptual Information

If you include steps in a concept topic that walk users through a task, you're not writing a concept topic. Move procedural, "how-to" information to a task topic. Also, if you've titled your task topics consistently by using verb-based or "how-to" titles, users won't expect to find a procedure in a concept topic with a noun-based title. Therefore, if you want users to find the right information, separate procedural information from conceptual information.

You can, however, use concept topics to describe processes, such as how coffee beans are cultivated or how data is sent through fiberoptic cables. In these topics, you can use a numbered list to outline the process. Also an effective image, such as a flow diagram or illustration.

Concept Components and DITA Elements

The DITA elements that you use in concept topics aren't specifically designed for conceptual material. Rather, you use common elements such as <p>, <ul>, and <section> elements to describe conceptual information.

You can use just about any DITA element in a concept topic except for elements specifically designed for task and reference topics. However, be careful that the conceptual information doesn't turn into a procedure or provide a long table of reference information.

Table 3.1 describes common elements used in concept topics.

Table 3.1 Common DITA XML Elements in DITA Concept Topics

DITA Element	Description
<title>	Provides a topic or section title
<shortdesc>	Introduces the concept.
<conbody>	Contains more description of the concept where you can add sections, paragraphs, and other elements
<section>	Contains a subsection that you can use to organize the conceptual information
<sl>	Displays a list of short or simple items
<ul>	Displays content as an unordered bulleted list
<dl>	Displays a list of terms or short concepts and their definitions
<fig> and <image>	Provides a figure and caption so that you can insert graphics
<term>	Highlights new terms

Titling the Concept Topic: <title>

Distinguish concept topics from task topics by using noun-based titles for concept topics and verb-based titles for task topics. For example, for a task topic title, use "Create user roles" or "Configure user roles." For a concept topic title, use "User roles" or "Role-based access."

Be sure to title topics consistently so that users know what type of information to expect in every topic. For more information about titles for topics, see the "Concept Topics" in *The IBM Style Guide* by De Respinis et al.

Introducing the Concept Topic: <shortdesc>

The concept topic short description should capture the main point of the topic. The introduction can start by:

- Defining a broad term or idea
- Describing the benefits of the tool, solution, part, or component
- Outlining a process

For example, in a topic called "Espresso beverages," your short description might start by defining espresso: "Espresso, sometimes called *caffe espresso* or *expresso*, is a concentrated beverage that is made by forcing pressurized water through finely ground coffee."

As another example, in a topic called "Web services for search engine systems," you might briefly describe the benefit of using web services in the short description: "To take advantage of open source technologies, you can use web services, which are XML-based applications, to configure your search application environment."

As always, keep short descriptions short. Use the <conbody> element to expand on the main point in the short description.

For more information about writing effective short descriptions, see the section "Concept Topic Short Descriptions" in Chapter 5, "Short Descriptions."

Writing the Concept: <conbody>

The main component of a concept topic is the concept body, or <conbody> element. You can use paragraphs, sections, images, examples, and other common elements in the <conbody> element.

After you briefly describe the main point of the concept in the <shortdesc> element, use the <p> element and other elements inside the <conbody> element to expand your discussion. For example, the concept topic called "Nuclear fusion as a power source for the Exprezzoh 9000N" describes how nuclear fusion creates energy, as shown in Figure 3.1.

Organizing the Concept: <section>

Use <section> elements to help users navigate and scan your content by organizing content into logical groups and providing headings, or section titles. Add titles to sections by using the <title> element.

Break up concept topics with sections and section titles in these situations:

- If the information in each section isn't long enough to be in its own topic.
- The information wouldn't make sense in its own topic.
- If you expect users to view the information together.

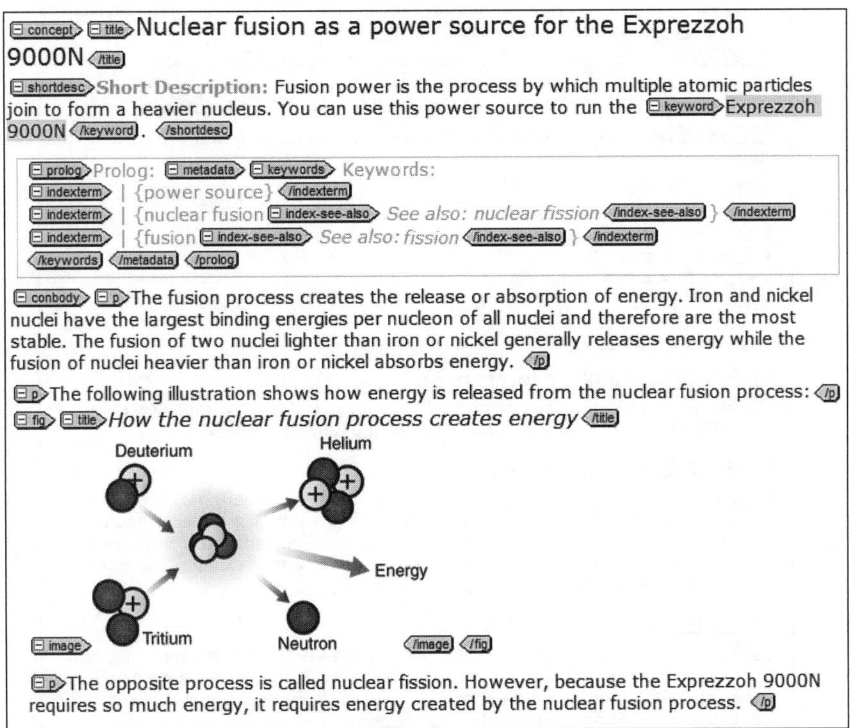

Figure 3.1 A concept topic that describes nuclear fusion as an energy source.

For example, in Figure 3.2 the concept topic called "Espresso beverages" describes common espresso drinks but separates those descriptions, using <section> and <p> elements.

Although the <title> element is optional for a <section> element, you should add titles in sections, which look like secondary headings in the output.

WATCH OUT Ensure that you don't overuse sections in concept topics. Cover one main concept per topic. Don't let a concept topic go on for five pages by trying to cover too much.

Adding Lists: <ol>, <ul>, <sl>, and <dl>

You can make your concept topics easy to scan if you use ordered, unordered, simple lists, or definition lists (<ol>, <ul>, <sl>, and <dl> elements, respectively). Limit most lists to seven items or fewer if the items are long, and nine items or fewer if the items are short.

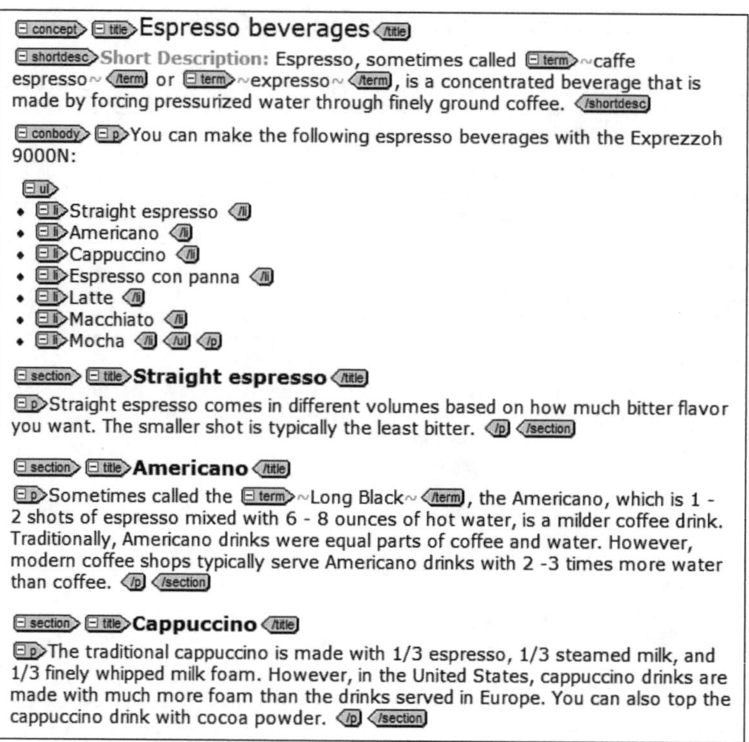

Figure 3.2 A concept topic with <section> and <p> elements.

Use Unordered Lists for Short, One-Part Items

If the content in your list item is more than one or two full sentences, consider presenting the information in another way, such as in a <section> or <dl> element. Figure 3.3 shows an unordered list of espresso beverages.

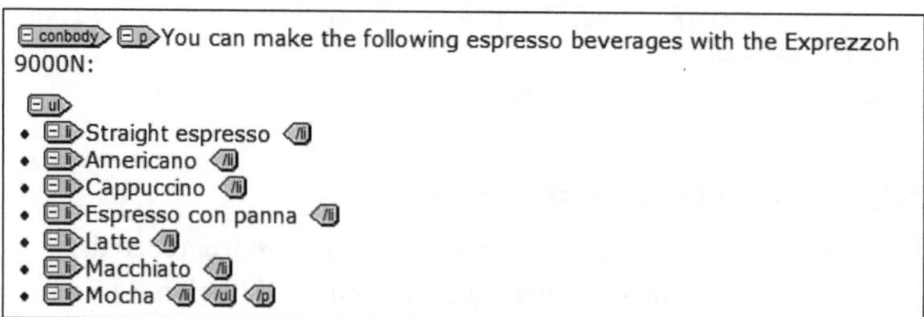

Figure 3.3 A typical unordered list created with the <ul> element.

Use Ordered Lists to Describe Workflows, Rankings, or High-Level Tasks That Aren't Specific Steps

You can use ordered lists to describe processes or other information that entail a chronological progression. Ensure that an ordered list isn't a user task disguised as a process. If you want users to follow a procedure, write a task topic.

The example topic in Figure 3.4 correctly uses an ordered list to describe the process of how cats manage to land on their feet. Describing this process in an ordered list is much easier to understand than having to read it in a paragraph.

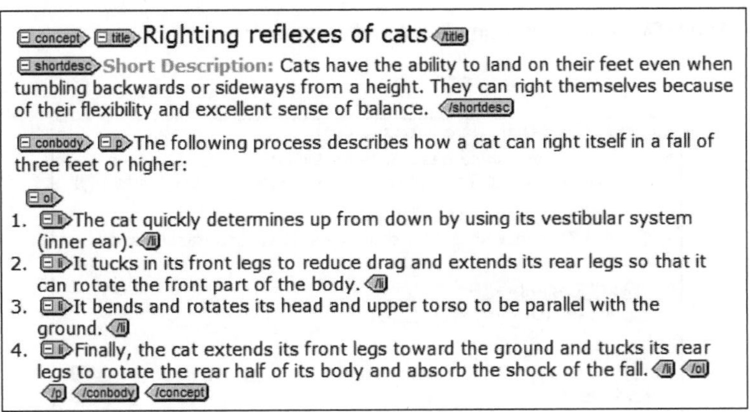

Figure 3.4 An ordered list created with the <ul> element that describes a process.

Use the Simple Lists (<sl> element) When You Don't Want to Show Bullets in the Output

If you use the simple list, limit the list items to short terms or phrases that are easy to scan without the visual aid of bullets.

Figure 3.5 shows a topic that uses a simple list. In HTML output, no bullets are shown.

Use Definition Lists (<dl> element) for Two-Part List Items

Use definition lists (<dl> element) for lists of terms and their corresponding definitions or descriptions. Figure 3.6 shows how a <dl> element describes different types of espresso drinks.

Be sure to use the definition list as it was intended: to define terms, objects, or short ideas. In DITA, you can't insert a <section> element inside another <section> element. Therefore, avoid using the <dl> element to create more levels of headings. If you think that your content needs three, four, or five levels of headings, you probably need to separate your information into two or more concept topics.

Caramel mocha latte

A caramel mocha latte is a sweet espresso drink that blends the chocolate of a mocha with caramel flavoring.

To make a caramel mocha latte, you need these ingredients:

2 ounces of caramel syrup
2 ounces of chocolate syrup
8 ounces of steamed milk
2-ounce shot of espresso

Figure 3.5 The HTML output of a simple list (<sl> element).

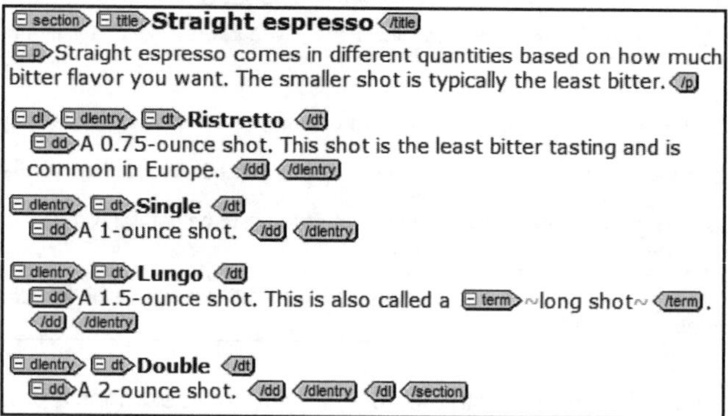

Figure 3.6 A concept topic that uses a definition list (<dl> element).

Including Graphics: <fig>, <title>, and <image>

You know the old adage about how many words you can get for a good picture. Nearly all readers prefer to view an illustration if it helps them avoid reading text. Include illustrations, diagrams, and other graphics to support conceptual information.

Figure 3.7 shows a photograph in a topic that describes how espresso machines work.

Highlighting New Terms: <term>

Use the <term> element to mark up new terms that are introduced and defined in concept topics. Use the <term> element only for the first or most prominent occurrence of the term. By default, content in the <term> element is italicized in the output.

For example, in a topic called "Cat vision," you can use the <term> element for the term *tapetum lucidem*, as shown in Figure 3.8.

Also, you might use the <term> element in more complex ways. For example, you can modify your CSS file to display term definitions when users hover over the term in HTML output.

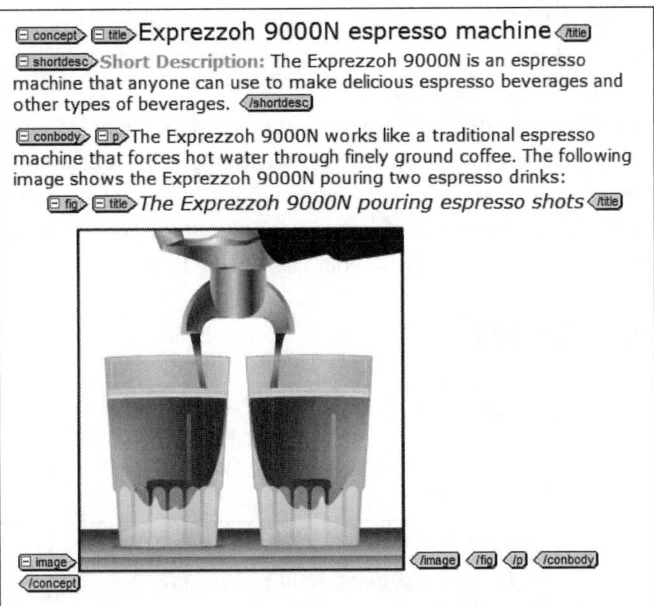

Figure 3.7 An image in a concept topic that describes espresso drinks.

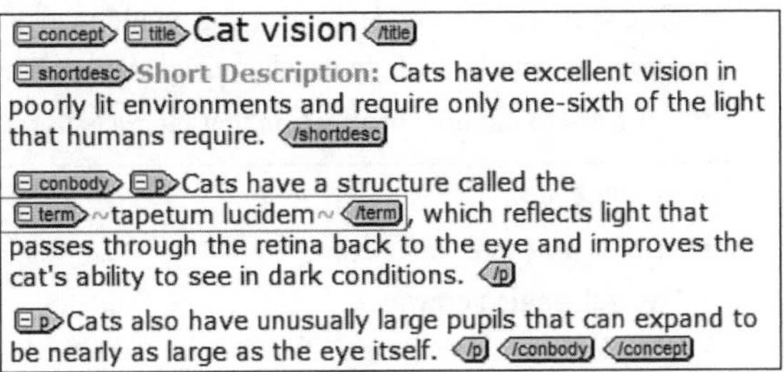

Figure 3.8 A concept topic that uses the <term> element for the term *tapetum lucidem.*

To Wrap Up

You build useful task-oriented information not only with good task topics, but also with good concept topics. To accomplish their goals, your users need clear, direct procedures, but they might also need more extensive conceptual information that can't be adequately described in the context of a task topic. You probably couldn't describe to novice users how to use a GPS device until they understand concepts such as waypoints, longitude, and latitude.

By separating conceptual information from tasks, you also make it easier for users to find only what they need. For example, novice users are more likely to need conceptual information than expert users. Those expert users also won't be bogged down by having to dig through conceptual information to find a task. They also won't need to skip over lengthy conceptual information in a task topic. Moreover, separating conceptual information from tasks provides more opportunities for you to reuse concept topics.

Organize conceptual information by using sections and lists. Don't forget to add diagrams, illustrations, and other graphics to enliven your information and to make it more informative.

And lastly, remember to write concept topics only as they're needed to help users get through their tasks. Good task-oriented information often requires more than just task topics, but don't create concept topics unless you know your users need them.

Concept Topic Checklist

Guideline	Description
Describe one concept per topic.	If the conceptual information is longer than one or two pages, break it up into smaller topics.
Use concept topics appropriately.	Create concept topics to support tasks. Use concept topics to: • Describe a system, product, or solution. • Provide a process overview or background information. • Introduce tools and technology. • Explain features, components, characteristics, restrictions, or capabilities. • Define terms in more detail than you would in a glossary. • Describe benefits or help users to make choices between options. Be sure not to include information that instructs users how to do something.
Use noun-based phrases for titles.	For example, use: • "Roles for espresso machine users" • "Search engine crawlers" • "Nuclear fusion overview"
Create effective concept topic short descriptions.	You can introduce the concept in the <shortdesc> element by: • Defining the object, component, feature, or technology • Describing benefits or importance of the object • Outlining a process
Organize the concept and break up dense text.	Use sections with titles, unordered lists, definition lists, and, in some cases, ordered lists to organize conceptual information.
Add images to describe the concept.	Use <fig> and <image> elements to insert images.
Use the <term> element appropriately.	Use the <term> element for termsyou describe for the first time.

CHAPTER 4

Reference Topics

When you write task-oriented technical content, you typically start by writing task topics that focus on user goals. Then, you write concept topics to explain ideas that users need to understand to complete the tasks. However, task and concept topics aren't always enough: You often need to include reference information to further support those tasks.

For example, if you want to describe how to change the oil in your car, you need to write the instructions in a task topic. But users can't finish that task unless they see a list of parts and tools needed for the job. You describe those parts and tools in a reference topic.

A *reference topic* is a collection of facts. Those facts might be a list or description of parts, commands, application programming interfaces (APIs), book titles, car models, animal species, bicycle tire sizes, or any other object that you can organize into a table or list.

Follow these guidelines when you write reference topics:

- Describe one type of reference material per topic.
- Organize reference information logically.
- Format reference information consistently.

Describe One Type of Reference Material per Topic

Make your content more usable and retrievable by limiting the type of reference information that you include in each topic. For example, most motorcycle shop manuals list the components of the bike by area: body, suspension, engine, and electrical system. If users want to find a wiring diagram, they simply look in the reference topic that describes the electrical system.

As a general guideline, if the reference information is short, say, 1 to 2 pages, include all the commands, features, or objects in one topic. If the reference information is longer than about two pages, break up the information into multiple tables or multiple topics.

Sometimes, it's more difficult to decide whether to put all the reference information in one topic or in multiple topics. For example, should you add all the available API methods for a software product in one topic, or should you describe each method in a separate topic?

To help you decide, consider how the reference information might be reused, organized, or searched, and whether that information might expand later. Will that short referenced item that you're describing require its own topic a year from now? Reference topics often start small, but sometimes they can grow like one of those overfed cats you see in Internet videos. You might also consider how the information is typically presented in documentation for your industry.

For example, in the case of the API methods, you can put each method in its own topic because that topic can be more easily reused, reorganized, or removed. Also, individual topics are easier to find from the table of contents, from indexes, or from searches. The disadvantage to creating separate topics for short reference information is that you must manage more files.

Organize Reference Information Effectively

Reference topics are designed for speed: Users want to find some tidbit of information quickly and move on. Use DITA elements that provide visual cues, such as lists, tables, and sections, to make the reference items easy to find.

Group the reference items in tables or lists in some logical way. For example, group the commands for setting up user roles and permissions in one table and then group the commands for monitoring the power system in another table.

After you group the items in a table or list, present those items in some logical and consistent order. Most users look for information by following some organizing principle. For example, you can arrange reference information alphabetically, chronologically, spatially, or in some other pattern that makes sense to users.

Format Reference Information Consistently

Use consistent formatting in topics that contain similar information. For example, if you create several reference topics that have tables of hardware parts, use the same formatting for the tables, use the same introduction to the tables, and provide the same amount of information. Even small inconsistencies are distracting and potentially confusing.

Reference Components and DITA Elements

Reference topics support nearly all the elements allowed in concept topics but support additional elements that are unique to reference topics, such as the special section in reference topics for syntax diagrams (<refsyn> element).

Table 4.1 describes common elements used in reference topics.

Table 4.1 Common DITA XML Elements in DITA Reference Topics

DITA Element	Description
<title>	Provides a topic or section title
<shortdesc>	Introduces the task
<refbody>	Contains the body of the topic
<section>	Organizes content in a topic into sections
<table>	Contains table content
<ul>	Displays content as an unordered bulleted list
<dl>	Displays a list of terms or phrases and their definitions
<example>	Contains example content that explains or supports the topic
<parml>	Displays parameters in a format similar to a definition list
<properties>	Lists properties in a table by type, value, and description
<simpletable>	Provides a table that doesn't require a title
<refsyn>	Contains a syntax diagram

Titling the Reference topic: <title>

Although you can't always distinguish concept titles from reference titles just by looking at the grammatical structure of the title, at least distinguish reference titles from task titles by using a noun phrase in the reference topic title. For example, use titles such as:

- "DITA XML elements"
- "Database monitoring commands"
- "Keyboard shortcut keys"
- "650cc engine specifications"
- "Required software for the ProStarEditor"

Even though concept and reference topic titles are noun-based, the content of the title should give users a hint that the topic describes conceptual or reference information.

For more information about titles for topics, see the section "Reference Topics" in Chapter 7 in *The IBM Style Guide* by De Respinis et al.

Introducing the Reference Information: <shortdesc>

Like all other topics, reference topics need effective short descriptions. In the short description, describe what the referenced item does, is used for, or its benefits.

For example, the short description in Figure 4.1 explains what predefined roles are used for and how using those roles can help users complete their tasks:

Avoid long introductions or discussions in reference topics. Your users probably already understand the concepts behind the reference information.

For example, if you're describing the commands for a web application, you shouldn't need to describe what a web application is or how it works. That information belongs in a concept topic.

For more information about writing effective short descriptions, see Chapter 5, "Short Descriptions."

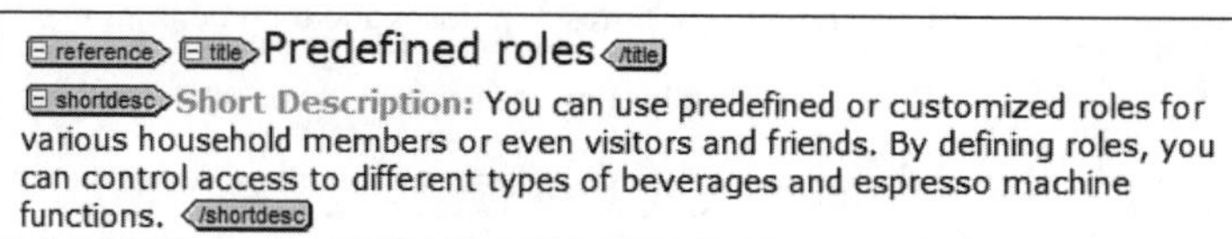

Figure 4.1 A short description for a reference topic that describes the main point of the reference information.

Organizing the Reference Information: <section>

After you briefly introduce the reference topic in the <shortdesc> element, use <section> elements in the <refbody> element to organize the information into logical sections.

For example, in a command reference topic for a software product, you might include sections with titles such as Description, Environment, Syntax, and Example. Users can scan information more quickly if a topic is organized into titled sections.

The reference topic in Figure 4.2 uses sections to organize separate tables.

BEST PRACTICE Although the <section> element allows more than one <title> element, use only one <title> element per section and place the <title> element as the first nested element in the section.

Not all <section> elements require a title. The reference topic requires a <section> element in the <refbody> element. However, if you have only one type of reference information and therefore only one section or that information is short, don't include a section title as shown in Figure 4.3.

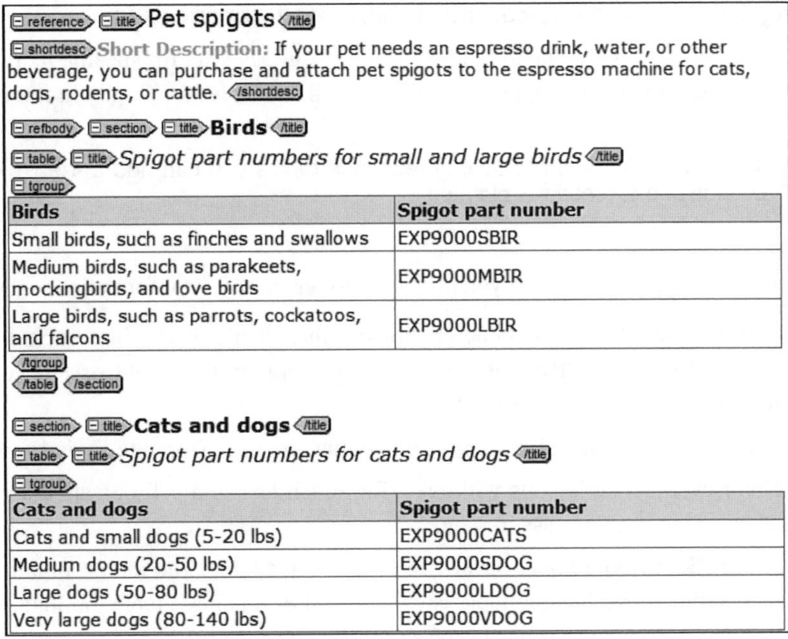

<reference> <title>Pet spigots</title>

<shortdesc>Short Description: If your pet needs an espresso drink, water, or other beverage, you can purchase and attach pet spigots to the espresso machine for cats, dogs, rodents, or cattle. </shortdesc>

<refbody> <section> <title>**Birds**</title>

<table> <title>*Spigot part numbers for small and large birds*</title>

<tgroup>

Birds	Spigot part number
Small birds, such as finches and swallows	EXP9000SBIR
Medium birds, such as parakeets, mockingbirds, and love birds	EXP9000MBIR
Large birds, such as parrots, cockatoos, and falcons	EXP9000LBIR

</tgroup>

</table> </section>

<section> <title>**Cats and dogs**</title>

<table> <title>*Spigot part numbers for cats and dogs*</title>

<tgroup>

Cats and dogs	Spigot part number
Cats and small dogs (5-20 lbs)	EXP9000CATS
Medium dogs (20-50 lbs)	EXP9000SDOG
Large dogs (50-80 lbs)	EXP9000LDOG
Very large dogs (80-140 lbs)	EXP9000VDOG

Figure 4.2 <section> elements that are used to organize content in a reference topic.

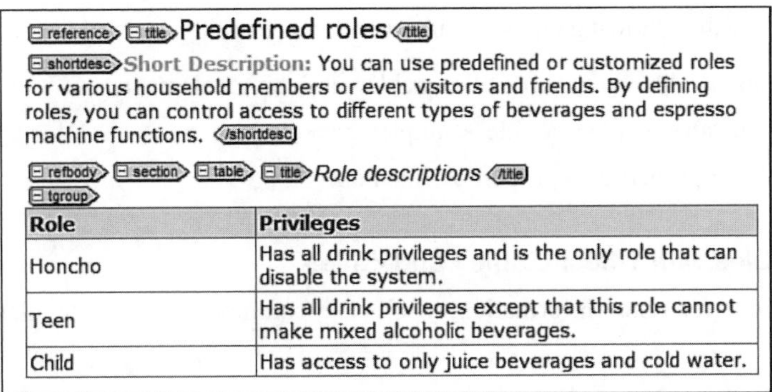

<reference> <title>Predefined roles</title>

<shortdesc>Short Description: You can use predefined or customized roles for various household members or even visitors and friends. By defining roles, you can control access to different types of beverages and espresso machine functions. </shortdesc>

<refbody> <section> <table> <title>*Role descriptions* </title>

<tgroup>

Role	Privileges
Honcho	Has all drink privileges and is the only role that can disable the system.
Teen	Has all drink privileges except that this role cannot make mixed alcoholic beverages.
Child	Has access to only juice beverages and cold water.

Figure 4.3 A short reference topic with only one section and no section title.

Creating Tables: <table>, <simpletable>, and <properties>

Tables provide an effective visual aid that helps users quickly scan for specific items. Tables are much easier to scan than paragraphs. Therefore, you'll use tables frequently in reference topics.

TIP Unlike concept topics, in reference topics you can add a <table> element outside of the <section> element.

Follow these guidelines when you use tables to organize your information:

- Use specific table titles (table captions) rather than generic titles such as "Table 1," "Table of parts," or "Parts list." Instead, use "Parts for the 1964 Corvette" or "Software requirements for the XBData Module."
- Use specific and logical headings for columns, such as "Corvette engine part number."
- Don't overload table cells with text. Too much text in a cell defeats the purpose of a table, which must be easy to scan.
- Avoid adding more than four or five columns in a table. Too many columns are just as ineffective as overloading table cells with text or making a table too long. Long tables that have too many columns or rows are difficult to scan.
- Avoid making the last column of a table a catchall for comments, examples, and descriptions. Separate this information into separate columns. For example, create a separate row for examples.

You can use the following types of tables in reference topics:

- A common table with a table title: <table> and <title> elements
- A simple table that has no title: <simpletable> element
- A table for properties: <properties> element

Common Tables with Titles: <table> and <title>

For most of your reference information, you'll use common tables with titles as shown in Figure 4.4.

Simple Tables with No Titles: <simpletable>

In some cases, you might create a short reference topic with just a topic title, short description, and one table. If the topic title and short description already accurately describe the content of the table, use the <simpletable> so that you don't need to include a table caption. By not including a table title, you avoid repeating information, such as repeating the topic title in the table title.

For example, the topic in Figure 4.5 uses a <simpletable> element. A table title, such as "Decorative and functional accessories for the espresso machine," is unnecessary because that's the title of the topic, and the topic contains no other sections or tables.

Hardware requirements

Hardware	Description
Memory and hard disk space	• 2 GB memory • 4 GHz processor • 400 GB disk space
Fusion reactor linkage	ExprezzohPower Station: provides the linkage from external power cables that go to your home from the reactor plant to the Exprezzoh 9000N.
Optional: MP3 players	• PlayIt 3.0 or later • WhalePod 9.1 or later • MP Fun 2012 or 2013
Cables and connectors	• ConnectoSoft cable: connects your computer to the Exprezzoh 9000N • ConnectoNuk cable: connects the Exprezzoh 9000N to the fusion reactor ExPower Station in your home. • Optional: Pet spigot connector.
Surveillance hardware	• GeigerFind 2012 Monitoring Dashboard: monitors your home for radioactivity • Optional: Web cam to watch your pets stare at or play with the swirling water

Figure 4.4 A common table that describes hardware requirements.

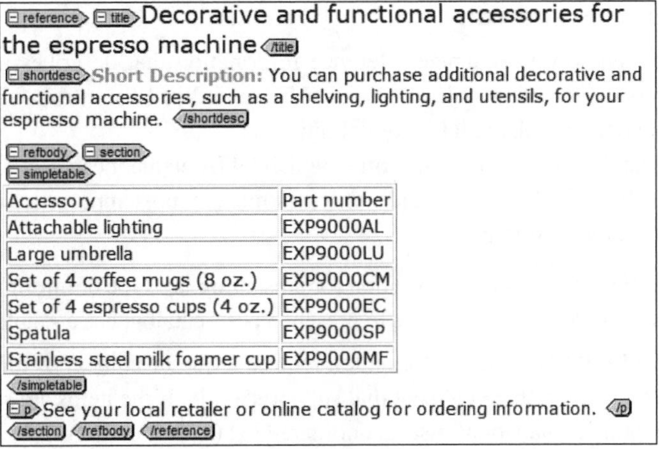

Decorative and functional accessories for the espresso machine

Short Description: You can purchase additional decorative and functional accessories, such as a shelving, lighting, and utensils, for your espresso machine.

Accessory	Part number
Attachable lighting	EXP9000AL
Large umbrella	EXP9000LU
Set of 4 coffee mugs (8 oz.)	EXP9000CM
Set of 4 espresso cups (4 oz.)	EXP9000EC
Spatula	EXP9000SP
Stainless steel milk foamer cup	EXP9000MF

See your local retailer or online catalog for ordering information.

Figure 4.5 Reference information in a <simpletable> element.

Properties Table: <properties>

Another type of table that you can use is the properties table. Use the <properties> element, which is valid only in reference topics, to create a table that lists a type, value, and description for properties.

For example, if you document how to design HTML pages, you might create a reference topic called “Cascading style sheet properties for body text” that includes a properties table describing the different cascading style sheet (CSS) properties for fonts. The table could describe font properties, such as family, style, weight, or size as shown in Figure 4.6.

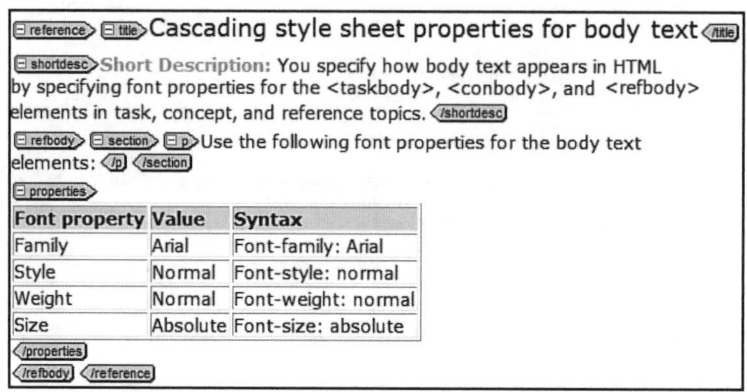

Cascading style sheet properties for body text

Short Description: You specify how body text appears in HTML by specifying font properties for the <taskbody>, <conbody>, and <refbody> elements in task, concept, and reference topics.

Use the following font properties for the body text elements:

Font property	Value	Syntax
Family	Arial	Font-family: Arial
Style	Normal	Font-style: normal
Weight	Normal	Font-weight: normal
Size	Absolute	Font-size: absolute

Figure 4.6 A <properties> element that creates a three-column table.

Adding Lists: <ul> and <dl>

Lists are a common way to organize reference information that describes parts, ingredients, requirements, models, or other objects. Avoid using unordered lists (<ul> element) if the list items have two parts. Instead, use tables or definition lists.

In a reference topic, the question is often which DITA element can best structure the information: tables or lists? To help you decide which element is most appropriate for your information, answer the following questions:

- **Convention:** What is the convention for presenting this type of information? If it’s common in the industry to use tables to present a parts catalog, use a <table> element.
- **Retrievability:** Does this information require a title or number to improve retrievability? If so, use a <table> element that supports a title. If the items are short because they don’t require explanations, use an unordered list (<ul> element).
- **Complexity:** Does the information have multiple parts, such as a model and its description? If the information has multiple parts, the content might best be organized as a table with multiple columns. Tables are typically easier to scan than lists, especially if the list items are long. If the two-part item is a term and its definition, use a definition list.

Creating Syntax Diagrams: <refsyn> and <syntaxdiagram>

Use the <syntaxdiagram> element inside a <refsyn> element to structure syntax diagrams that illustrate the syntax of commands, utilities, or other items that have a specific grammar.

The topic in Figure 4.7 includes a syntax diagram that shows command options and the order that those options must be entered for the LOAD command.

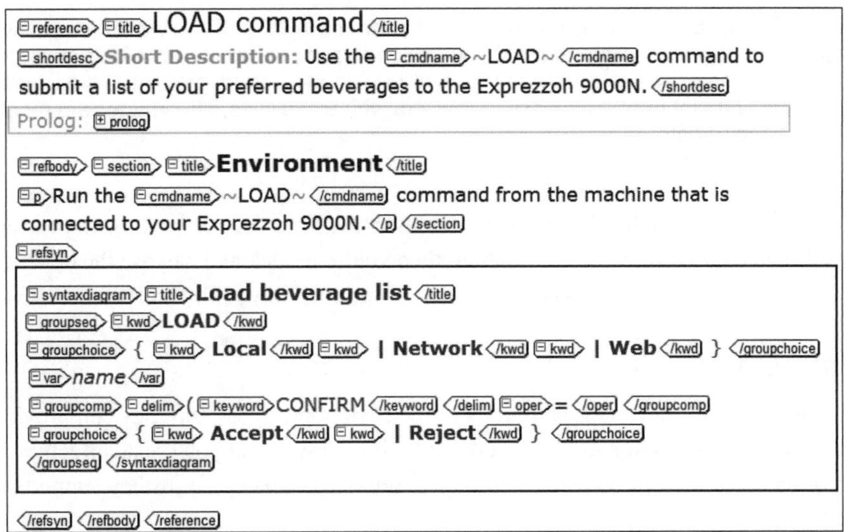

Figure 4.7 A syntax diagram in a reference topic.

Entering DITA markup for syntax diagrams is a skill in itself, so learning to create them can take some time.

The DITA Open Toolkit creates syntax diagrams in output as the Backus-Naur form. Backus-Naur (BNF) is a common notation for computer science but might not best serve your users. The following example shows the Backus-Naur output for the DITA syntax markup from Figure 4.7:

```
LOAD {Local | Network | Web} name [ ( CONFIRM={Accept | Reject}]
```

If you want to represent syntax diagrams in an alternative form, such as railroad diagrams, you must change your transform to produce a different style in the output.

Figure 4.8 shows how Railroad diagrams provide a visual representation of the syntax of a command, utility, or language.

```
>>-required_item--+--------------------------------------+-----><
                  |                 .-default_choice--.  |
                  +-optional_choice--+-required_choice-+-+
                  |                 '-required_choice-'  |
                  '-optional_choice----------------------'
```

Figure 4.8 A railroad diagram that represents command syntax.

TIP Syntax diagrams can be difficult to create in DITA. If you want to avoid syntax diagrams, we recommend that you use code samples as a way to demonstrate the proper grammar of a command. Avoid using images to present syntax because images might not be accessible to users with disabilities, and images are more difficult to update.

Consider which graphical representation your users are accustomed to. If your audience understands and expects Backus-Naur output, then you're in luck and can use the default output. Otherwise, you must modify the way that syntax is presented in your information.

To Wrap Up

Reference topics aren't the most glamorous of the basic topic types, but they can provide quick access to valuable information. As with concept topics, be sure that you create effective task-oriented information, which means that you include reference topics only if they support real user goals and tasks.

By now, you've memorized the mantra to describe only one type of reference information per topic. However, deciding what content should be included in a reference topic can be problematic. For example, should you describe all commands or APIs for an application in one topic or separate the information into multiple topics? Creating separate topics, even for short content, is generally more useful for both you and your users. However, you might need to manage many more files.

For most reference information, avoid creating topics that are more than two pages. However, if it makes sense to include all the parts and part numbers, models, API methods, ingredients, or commands in one topic, then do so.

Also, be sure not to write reference topics that users will likely never need, such as descriptions of interface toolbar icons for software products. Don't "paper" the product by adding unnecessary topics that users will never need.

Organize reference information for quick retrieval and easy scanning by using sections, tables, and lists, and by organizing that information logically and consistently. For example, use the same formatting for tables so that users aren't distracted or confused by inconsistencies.

Effective reference topics are essential in nearly all technical information. Clear, organized reference topics can help users use your product more efficiently to get their jobs done.

Reference Topic Checklist

Guideline	Description
Use noun-based phrases for titles.	For example, use: • "Power management commands" • "Mountain bike models" • "Domestic dog breeds"
Create effective reference topic short descriptions.	In the short description, describe what the referenced item does, is used for, or its benefits.
Describe one type of reference information per topic.	Be consistent about how much and what type of content you add to your reference topics. For example, decide whether to describe some or all related items in one topic or describe one item per topic.
Use sections to organize the reference information.	To make your reference information easier to scan, divide your content into sections and use one only title per section. Omit the section title if the topic is short and has only one section.
Create consistent reference information.	Use similar formatting, organization, titles, and phrasing for tables, lists, and sections.
Use the correct type of table.	For example, use: • The <table> element for tables that require a title • The <simpletable> element for short tables that don't require a title • The <properties> element for tables that describe types, values, and descriptions for each item

Guideline	Description
Create effective tables.	Follow these guidelines when you use tables: • Use specific titles (table captions) rather than generic titles, such as "Table 1," "Table of parts," or "Parts list." Instead, use "Parts for the 1964 Corvette" or "Software requirements for the XBData Module." • Use specific and logical headings for columns, such as "Corvette engine part number." • Don't overload table cells with text. Too much text in a cell defeats the purpose of a table, which must be easy to scan. • Avoid adding more than four or five columns in a table. Too many columns are just as ineffective as overloading table cells with text or making a table too long. Long tables that have too many columns or rows are difficult to scan. • Avoid making the last column of a table a catchall for comments, examples, and descriptions. Separate this information into separate columns. For example, create a separate row for examples.
Evaluate your needs for syntax diagrams.	Decide which output format you need for syntax diagrams. Remember that you and your team need to build the skills necessary to create syntax diagrams.

CHAPTER 5

Short Descriptions

The <shortdesc> element is perhaps the most versatile and yet most challenging element to write for because this element does more than just act as the first paragraph in a topic. The short description text can be seen in links, and in some cases, search engine results.

After you convert content to DITA or when you start writing your first DITA topics, make writing effective short descriptions a priority.

The <shortdesc> Element

The <shortdesc> element contains the first paragraph of a topic and is valid in several locations:

- Between the <title> and body elements of standard topics (concept, task, and reference)
- Inside the <abstract> element, which also goes between the <title> and body elements
- Inside a <topicref> element in a DITA map

How the Short Description Is Used

So, what's the big deal? This is one element among more than 400 DITA elements. Why is an entire chapter based on a single element?

The <shortdesc> element seems easy to understand, but knowing what to write in that element can be challenging. The <shortdesc> element is difficult to write for because it does more than provide the first paragraph of a topic. The short description text can be viewed in any of these contexts:

- The first paragraph of a topic
- Link previews for related or child topic links
- As abstracts for search engine results

First Paragraph in a Topic

The short description must succinctly describe the main theme or purpose of the topic. When users open a new topic and read the first sentence, they can quickly decide whether they've found the correct information.

BEST PRACTICE In technical documentation, avoid building up to a main point. Instead, write short descriptions that briefly summarize the main point of the topic.

Figure 5.1 shows a short description that describes the main point of the task topic and why installing the software is necessary.

task title Install the espresso machine software /title
shortdesc Short Description: After you connect your espresso machine to the fusion power, install the espresso machine software on your computer so that you can configure roles, create schedules, mix drinks, send alarms, monitor the power system, and manage other activities.
/shortdesc

Figure 5.1 An effective opening paragraph for a task topic in the <shortdesc> element.

In the output of the topic, the short description looks like a standard paragraph, as shown in Figure 5.2.

Install the espresso machine software

After you connect your espresso machine to the fusion power, install the espresso machine software on your computer so that you can configure roles, create schedules, mix drinks, send alarms, monitor the power system, and manage other activities.

Figure 5.2 The HTML output of the short description as the opening paragraph for a task topic.

Link Preview for Related or Child Links

DITA short descriptions can appear as link previews. Generated links can display not only the topic title but also the short description for that topic. The short description acts as a preview of the target topic content.

Link previews are generated automatically. When you build output of your DITA topics, the titles and short descriptions displayed in the links are pulled directly from the target topics. You don't create these links manually.

Figure 5.3 shows how users can hover over a hyperlink to preview the short description of a topic before clicking the link. It also shows how the short descriptions for child topics are pulled into the parent topic output.

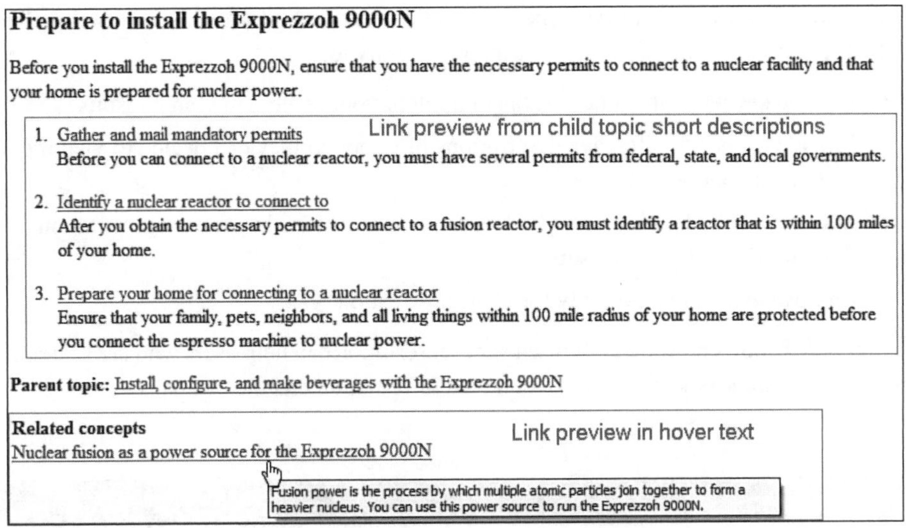

Figure 5.3 A short description of a topic that is displayed in HTML output when users hover the cursor over the link and as short descriptions of the child topics.

These low-maintenance linking features can improve the usability of your documentation. By reading the link preview, users can determine whether the topic contains the information that they want before they click the link. Therefore, your short description must describe the topic concisely and accurately without simply repeating the topic title.

Although the previous image shows the short description for the related link as hover text, you can also set up the link preview to display the title and short description as plain text.

BEST PRACTICE Use the default behavior to display the short description as hover help for links. This strategy supports the model of progressive disclosure, in which information is revealed to users only when they need it. If the topic title in the link text provides enough information, users don't need the short description to clutter their view. If users do need additional information, they can hover over the link to read the short description.

Search Engine Results

If you enable your online HTML content to be crawled by public search engines, the short description is displayed in the search results. How effectively users can locate information in your content might be the most important challenge to creating effective technical documentation. If your users can't find the information they need, it doesn't matter how well written the topics are.

One way that you can improve the retrievability of your information is to write effective short descriptions. The short description is an opportunity to:

- Provide key terms that improve the ranking of your content in search results
- Provide users with effective descriptions of topics so they can decide to visit the information on your website
- Save money by helping users find answers to questions in your documentation so they don't need to call product support
- Improve overall retrievability for your information

Figure 5.4 shows how a well-written short description can help users who are searching for information determine whether they've found the information that they want.

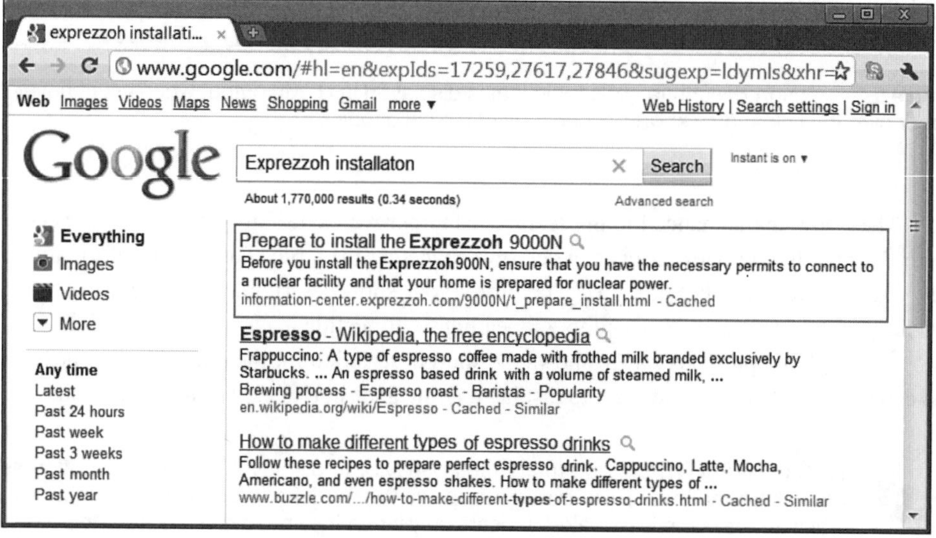

Figure 5.4 Short description text as shown in search engine results.

Search results that include short description text can also display in some HTML help systems, such as Eclipse information centers, as shown in Figure 5.5.

Guidelines for Writing Effective Short Descriptions

To write short descriptions that are effective in all the venues where the short description can be displayed, follow these guidelines.

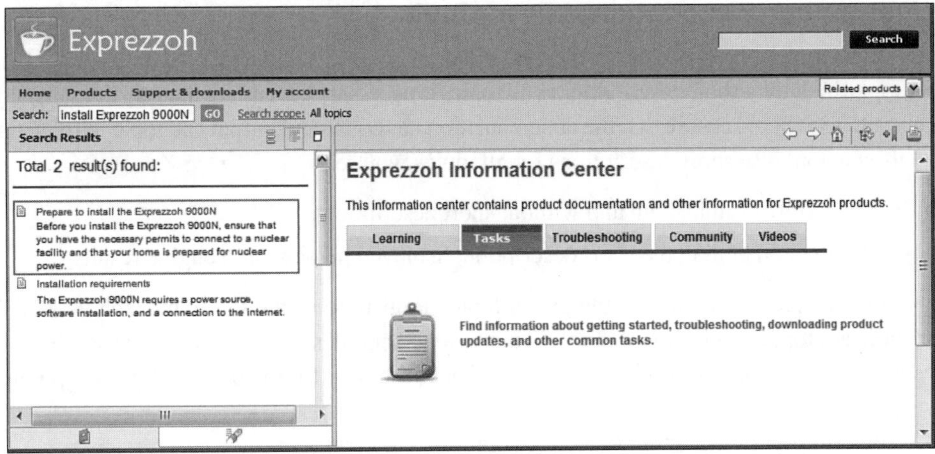

Figure 5.5 An Eclipse information center that shows short description text in search results.

Briefly State the Purpose of the Topic

The short description should describe the purpose or main point of the topic. An effective short description typically answers two questions:

- What is the topic about?
- Why do users care about or need the information in the topic?

Don't use the short description to build up to a point as shown in the following example:

Incorrect

```
<title>Mountain bike disc brakes and V-brakes</title>
<shortdesc>Disc brakes first appeared on higher end mountain
bikes in the mid 2000s and are now standard on most mountain
bikes.</shortdesc>
```

Correct

```
<title> Mountain bike disc brakes and V-brakes</title>
<shortdesc>Disc brakes have several advantages over tradi-
tional V-brakes. Disc brakes are better able to dissipate
heat in heavy braking conditions, and they perform better in
muddy and wet conditions.</shortdesc>
```

Include a Short Description in Every Topic

If you decide to use the <shortdesc> element, (and, of course, you should), you must include it in every topic. Including short descriptions in all topics is necessary because of the way links display in the output. You should ensure that the documentation has consistent output and topic introductions.

If you don't use short descriptions for all topics, you see:

- A mixture of links with and without short descriptions
- Content other than the short description displayed in the search engine results

For example, in Figure 5.6, the parent topic "Prepare to install the Exprezzoh 9000N" has four child task topics. Only two of the child topics have short descriptions, which causes the output to be inconsistent. Your users might even think that you simply forgot to add short descriptions.

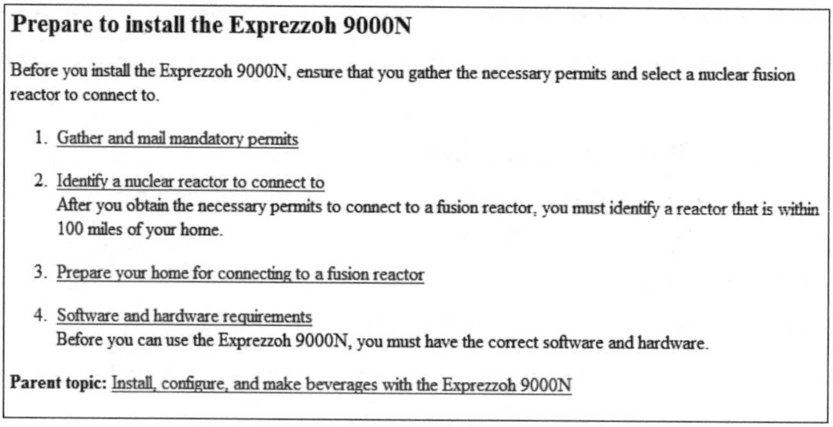

Prepare to install the Exprezzoh 9000N

Before you install the Exprezzoh 9000N, ensure that you gather the necessary permits and select a nuclear fusion reactor to connect to.

1. Gather and mail mandatory permits
2. Identify a nuclear reactor to connect to
   After you obtain the necessary permits to connect to a fusion reactor, you must identify a reactor that is within 100 miles of your home.
3. Prepare your home for connecting to a fusion reactor
4. Software and hardware requirements
   Before you can use the Exprezzoh 9000N, you must have the correct software and hardware.

Parent topic: Install, configure, and make beverages with the Exprezzoh 9000N

Figure 5.6 Short descriptions that are missing from links to task topics 1 and 3.

Missing short descriptions also affect search engine results. Users will see text from other parts of the topic that might be less effective at describing the main point of the topic. For example, the search results for an API reference topic that lacks a short description might show confusing or even nonsensical text, as shown in Figure 5.7.

TIP To ensure that all topics have short descriptions, make the <shortdesc> element a required element in topics or modify your transform settings to issue warnings for topics that are missing the <shortdesc> element.

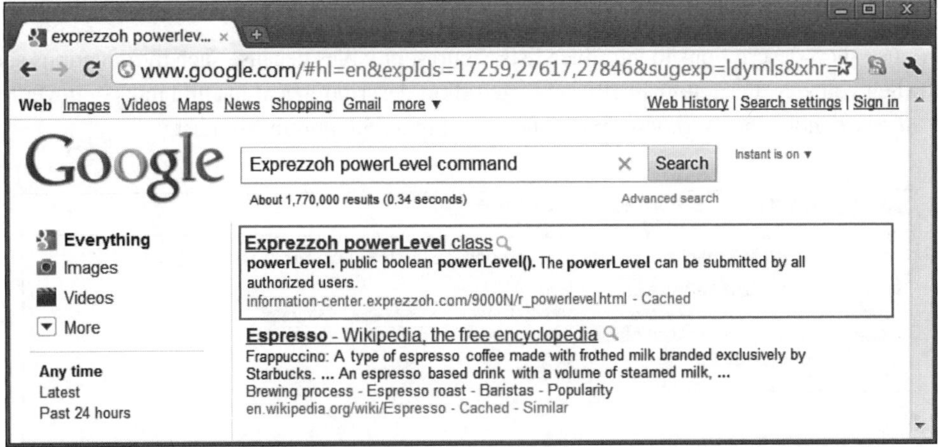

Figure 5.7 Confusing search results that show text from other parts of the topic because the topic has no short description.

Use Complete, Grammatical Sentences

For consistency across topics, ensure that your short descriptions are complete sentences and follow standard grammar, punctuation, and style rules. Incomplete sentences can look odd when other short descriptions are complete sentences. Also, complete sentences are easier for native and nonnative speakers to read and are easier to translate.

Incorrect

```
<title>Cappuccino</title>
<shortdesc>Contains 1/3 espresso, 1/3 steamed milk, and 1/3
finely whipped milk foam.</shortdesc>
```

Correct

```
<title>Cappuccino</title>
<shortdesc>The traditional cappuccino is made with 1/3
espresso, 1/3 steamed milk, and 1/3 finely whipped milk
foam.</shortdesc>
```

WATCH OUT Some writing teams might decide to use incomplete sentences for short reference topics, such as topics for API information. If you decide to use incomplete sentences in these cases, remember that your short descriptions will be inconsistent compared to other topics in your information set.

Don't Introduce Lists, Figures, or Tables

Don't use short descriptions to introduce lists, figures, or tables because such introductions seem odd when you view the automatically generated short description text in child links or hover text.

For example, the short description shown in Figure 5.8 introduces a list.

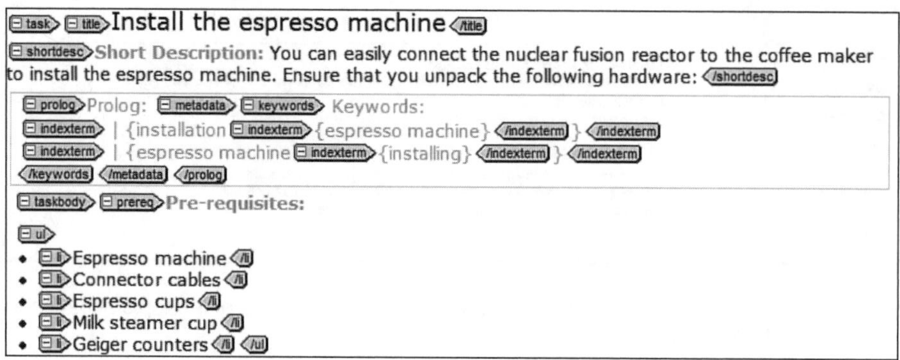

Figure 5.8 An ineffective short description that introduces a list.

A short description that introduces a list might seem fine when you first read the topic, but notice what happens when you build output, which generates child links. The short description for the topic "Install the espresso machine" ends with a colon, as shown in Figure 5.9.

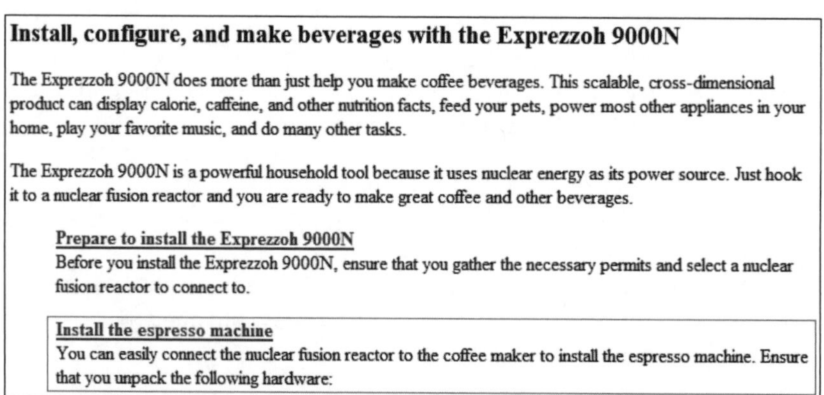

Install, configure, and make beverages with the Exprezzoh 9000N

The Exprezzoh 9000N does more than just help you make coffee beverages. This scalable, cross-dimensional product can display calorie, caffeine, and other nutrition facts, feed your pets, power most other appliances in your home, play your favorite music, and do many other tasks.

The Exprezzoh 9000N is a powerful household tool because it uses nuclear energy as its power source. Just hook it to a nuclear fusion reactor and you are ready to make great coffee and other beverages.

Prepare to install the Exprezzoh 9000N
Before you install the Exprezzoh 9000N, ensure that you gather the necessary permits and select a nuclear fusion reactor to connect to.

Install the espresso machine
You can easily connect the nuclear fusion reactor to the coffee maker to install the espresso machine. Ensure that you unpack the following hardware:

Figure 5.9 A short description as shown with a child link for a topic that introduces a list.

An introduction to a list in the short description also doesn't seem to make sense in hover text for a related link, as shown in Figure 5.10.

Parent topic: Install, configure, and make beverages with the Exprezzoh 9000N

Related concepts
Nuclear fusion as a power source for the Exprezzoh 9000N

Related tasks
Install the espresso machine

You can easily connect the nuclear fusion reactor to the coffee maker to install the espresso machine. Ensure that you unpack the following hardware:

Figure 5.10 A short description that introduces a list as seen in hover text for a link.

As seen in the link text, the short description for the topic "Install the espresso machine" tries to introduce some illusory list that no one can see. Your users might even think something is broken! To correct this problem, simply move the list introduction to the body element of the topic and rewrite the short description. Figure 5.11 shows how removing the list introduction from the short description improves the link preview in the output.

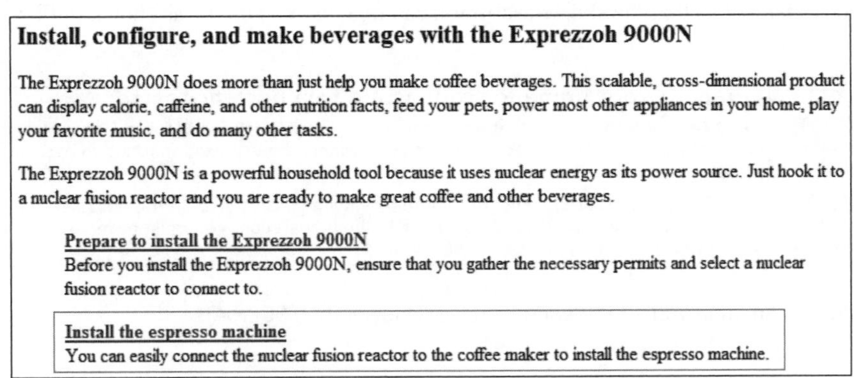

Install, configure, and make beverages with the Exprezzoh 9000N

The Exprezzoh 9000N does more than just help you make coffee beverages. This scalable, cross-dimensional product can display calorie, caffeine, and other nutrition facts, feed your pets, power most other appliances in your home, play your favorite music, and do many other tasks.

The Exprezzoh 9000N is a powerful household tool because it uses nuclear energy as its power source. Just hook it to a nuclear fusion reactor and you are ready to make great coffee and other beverages.

Prepare to install the Exprezzoh 9000N
Before you install the Exprezzoh 9000N, ensure that you gather the necessary permits and select a nuclear fusion reactor to connect to.

Install the espresso machine
You can easily connect the nuclear fusion reactor to the coffee maker to install the espresso machine.

Figure 5.11 An improved short description for a child link that doesn't introduce a list.

Keep Short Descriptions Short

Because of the different ways short description text can be displayed in output, you need to keep short descriptions short. Lengthy text in the <shortdesc> element makes it difficult to scan automatically generated child links and hover text for links. In rare cases, the short description can be up to 50 words, but keep most under 35 words.

Figure 5.12 shows a lengthy short description for the topic "Prepare to install the Exprezzoh 9000N" that not only takes up a great deal of space when displayed as a preview for a child topic, but the wordy description is also confusing.

Prepare to install the Exprezzoh 9000N

Before you install the Exprezzoh 9000N, ensure that you have the necessary permits to connect to the nuclear facility of your choice and that your home is prepared for nuclear power.

1. Gather and mail mandatory permits
   You can use the permit forms that are provided with the espresso machine. You must complete each form and have them signed by the appropriate official.
2. Identify a nuclear reactor to connect to
   After you obtain the necessary permits to connect to a fusion reactor, you must identify a reactor that is within 100 miles of your home.
3. Prepare your home for connecting to a nuclear reactor
   Ensure that your family, pets, neighbors, and all living things within 100 mile radius of your home are protected before you connect the espresso machine to nuclear power. You must prepare your home by setting up Geiger counters, radiation containment vessels, and other hardware. To complete the preparation, you must have your home inspected by nuclear facility inspectors.

Figure 5.12 A long short description that is difficult to scan.

To break up long short descriptions, you can simply move much of the short description text to the body element so that the short description can live up to its name, as shown in Figure 5.13.

<task> <title>Prepare your home for connecting to a nuclear reactor</title>

<shortdesc>Short Description: Ensure that your family, pets, neighbors, and all living things within 100 mile radius of your home are protected before you connect the espresso machine to nuclear power. </shortdesc>

<taskbody> <context> <p>You must prepare your home by setting up Geiger counters, radiation containment vessels, and other hardware. To complete the preparation, you must have your home inspected by nuclear facility inspectors.</p>

Figure 5.13 An improved short description that is fewer than 35 words.

The short description in the child link for the topic "Prepare your home for connecting to a nuclear reactor" is now more effective, as shown in Figure 5.14.

Avoid Writing Short Descriptions That Are Self-Referential

No one likes a narcissist, so avoid short descriptions that start with language that makes the topic the focus of the short description. For example, avoid the following phrasing:

- "This topic describes...."
- "This concept covers...."
- "This information is about...."
- "The following chapter explains...."

These constructions take up valuable space without adding any useful information.

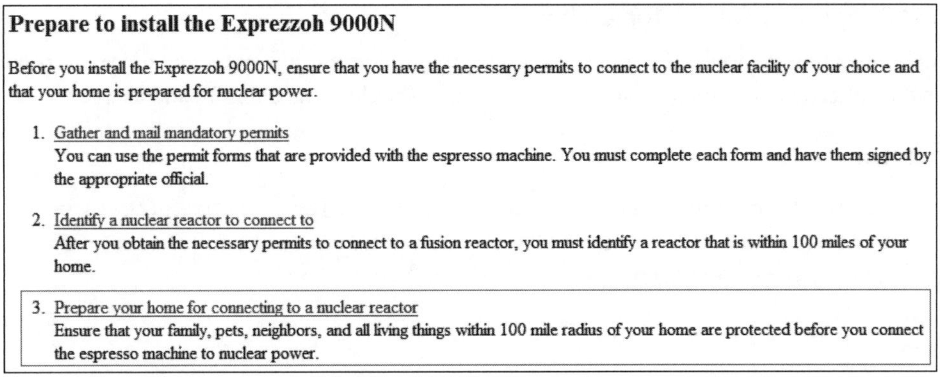

Prepare to install the Exprezzoh 9000N

Before you install the Exprezzoh 9000N, ensure that you have the necessary permits to connect to the nuclear facility of your choice and that your home is prepared for nuclear power.

1. Gather and mail mandatory permits
   You can use the permit forms that are provided with the espresso machine. You must complete each form and have them signed by the appropriate official.
2. Identify a nuclear reactor to connect to
   After you obtain the necessary permits to connect to a fusion reactor, you must identify a reactor that is within 100 miles of your home.
3. Prepare your home for connecting to a nuclear reactor
   Ensure that your family, pets, neighbors, and all living things within 100 mile radius of your home are protected before you connect the espresso machine to nuclear power.

Figure 5.14 An improved short description that appears in a child link.

Incorrect

```
<title>Mountain bike wheel comparison</title>
<shortdesc>This topic compares 26-inch mountain bike wheels
to 29-inch wheels.</shortdesc>
```

Correct

```
<title>Mountain bike wheel comparison</title>
<shortdesc>Deciding which wheel size to use for your mountain
bike depends on what environments you ride in. The 26-inch
wheel and the 29-inch wheel both have advantages and disad-
vantages depending on where and how you ride.</shortdesc>
```

Incorrect

```
<title>Coffee harvesting and processing</title>
<shortdesc>The following information describes how coffee
cherries are harvested and processed to become coffee beans
that you can use to make espresso drinks.</shortdesc>
```

Correct

```
<title>Coffee harvesting and processing</title>
<shortdesc>To create the coffee used in espresso drinks or
brewed coffee, the fruit of the coffee plant is harvested,
hulled, cleaned, aged, and finally ground.</shortdesc>
```

Don’t Simply Repeat the Topic Title

Simply repeating the topic title in the short description doesn’t give users any more information than they already know. Every sentence in your topic should be meaningful and useful.

Incorrect

```
<title>Start the system administration client on UNIX</title>
<shortdesc>You can start the system administration client on
UNIX systems.</shortdesc>
```

Correct

```
<title>Start the system administration client on UNIX</title>
<shortdesc>Start the system administration client on the web
application server so that you can manage your deployments.
Use the AdminClient.sh command to start the cli-
ent.</shortdesc>
```

Incorrect

```
<title>Radiological imaging technologies</title>
<shortdesc>Radiological imaging technologies are useful for
medical professionals.</shortdesc>
```

Correct

```
<title>Radiological imaging technologies</title>
<shortdesc>Physicians and other medical professionals use im-
aging technologies, such as radiography, MRI, fluoroscopy, CT
scans, and ultrasound, to diagnose disease and inju-
ries.</shortdesc>
```

Don’t Use Cross-References in the Short Description

Short descriptions are meant to describe the purpose and theme of the topic. If you include references, such as citations to other books, in the short description, you invite users to view other content before they read the current topic. That’s why DITA doesn’t allow you to insert an <xref> element in a <shortdesc> element.

BEST PRACTICE Avoid entering URLs as text in the <shortdesc> element. Instead, add links in more appropriate areas such as the <prereq> element or a relationship table.

Incorrect

```
<title>Install additional patches</title>
<shortdesc>After you install version 10, you must install the
latest patches for both the client and the server. Go to
www.xyzsoftware/version10/patch_city.com to find the latest
patches.</shortdesc>
```

Correct

```
<title>Install additional patches</title>
<shortdesc>After you install version 10, you must install
  the latest patches for both the client and the
  server.</shortdesc>
<taskbody><prereq><p>Go to <xref>www.xyzsoftware/version10/
patch_city.com</xref> to find the latest patches for version
10.</p></prereq></taskbody>
```

Short Descriptions for Task, Concept, and Reference Topics

In addition to the general guidelines for writing effective short descriptions, follow these additional guidelines for writing short descriptions for the three basic topic types: task, concept, and reference.

Task Topic Short Descriptions

To write effective short descriptions for task topics, answer one or more of the following questions:

- What does the task help users accomplish?
- What are the benefits of doing this task or why is the task important?
- When would users perform the task?
- What's involved in the task?
- Why do users need to perform this task?
- How does the task fit with other related tasks?

Focus on the Benefits or Importance of the Task

Describe why the task is important or beneficial. Your users know a procedure is coming, and they can tell what that task can help them to accomplish by reading the title. However, state why a task is useful or required. You might think the benefits are obvious, but describing *why* users must do a task can be helpful.

Incorrect

```
<title>Store excess fusion energy</title>
<shortdesc>This procedure shows you how to store excess
fusion energy.</shortdesc>
```

Correct

```
<title>Store excess fusion energy</title>
<shortdesc>To avoid radiation leaks, you must configure the
LeakNot storage device so that the excess fusion energy can
be safely stored.</shortdesc>
```

Incorrect

```
<title>Monitor power output</title>
<shortdesc>You can monitor the power output of your fusion
power system.</shortdesc>
```

Correct

```
<title>Monitor power output</title>
<shortdesc>Although the espresso machine power system runs
automatically, you must check the power output periodically
to ensure that the fusion energy is not overloading your
electrical circuits or whether you can divert excess power to
other appliances.</shortdesc>
```

Provide an Overview of the Procedure

Provide a quick overview of the steps or describe where and when to do the task so that users understand what the task helps them accomplish. For advanced users, the short description alone might provide enough information for them to do the task. Other users can follow the steps in the other part of the topic.

Incorrect

```
<title>Configure roles</title>
<shortdesc>You can configure roles and privileges for house-
hold members who use the espresso machine.</shortdesc>
```

Correct

```
<title>Configure roles</title>
<shortdesc>To configure roles, decide which members of your
household require administrator access. The administrator can
then set each role on the Manage Roles page from the dash-
board.</shortdesc>
```

Focus on Real Goals, Not Product Functions

Even though users often perform tasks in software interfaces, don't focus only on that interface, function, panel, window, menu, or wizard. Focus on the real user goal. The goal is never simply to use some window or tool in a product. You can mention the tool, window, or interface control, but do so only to orient the user.

For example, when you write task topics that describe how to do a task in a specific window or how to use a tool, mention the window or tool, but then briefly describe why the task is useful or how users can accomplish the goal.

Incorrect

```
<title>Change public display names</title>
<shortdesc>Use the Edit Profile page to change display names,
summaries, and contact information.</shortdesc>
```

Correct

```
<title>Change public display names</title>
<shortdesc>You can change the display name for your public
profile by using the Edit Profile page. Changing your display
name will not change URL that others use to access your
page.</shortdesc>
```

Incorrect

```
<title>Change service console settings</title>
<shortdesc>Use the CHANGESET command to change service con-
sole settings.</shortdesc>
```

Correct

```
<title>Change service console settings</title>
<shortdesc>If the service console graphical interface is in-
accessible because of an incompatible browser, you can change
```

```
the service console settings by using the CHANGESET command
from the command line.</shortdesc>
```

Provide Brief Conceptual Information

Although task topics focus on procedural information, you can also include some conceptual information in the <shortdesc> element to provide context for the task, such as when and where users do the task.

Incorrect

```
<title>Install the Exprezzoh software by using a response
file</title>
<shortdesc>Use the response file to install the espresso ma-
chine software.</shortdesc>
```

Correct

```
<title>Install the Exprezzoh software by using a response
file</title>
<shortdesc>A response file contains the values that you add
for your server environment. The installation program uses
those values in the file to install the Exprezzoh soft-
ware.</shortdesc>
```

If your conceptual information is longer than two sentences, you should move it to the <context> element or split it between the <shortdesc> and <context> elements. Keep most short descriptions under 35 words.

Show How Tasks Fit Together

In the task topic short description, help users to understand the relationship between the current task and other tasks. In some cases, you can mention the tasks that come before or after the current task in the short description. However, don't use language such as "In the next topic..." because your content will be difficult to reuse, and you might someday need to reorder the task topics.

Incorrect

```
<title>Connect to new devices</title>
<shortdesc>You might need to reset the machine so that the
espresso machine checks for any new devices. Using newly at-
tached accessories is covered in the next topic.</shortdesc>
```

Correct

```
<title>Connect to new devices</title>
<shortdesc>Before you can use new devices, you might need to
reset the espresso machine so that it checks for new de-
vices.</shortdesc>
```

Concept Topic Short Descriptions

To write effective short descriptions for concept topics, answer one or more of the following questions:

- What's the object, concept, or idea?
- Why should users care about this object, concept, or idea?

Briefly Define the Object or Idea

Define or introduce the feature, process, system, technology, or tool. However, don't use the short description to build up to a main point of the topic. The short description *is* the main point.

Incorrect

```
<title>Flashbacks</title>
<shortdesc>Different issues can affect your system. You might
  experience power outages, hardware failures, or software
  errors.</shortdesc>
<conbody><p>To preserve your system and have the ability to
revert to a specific state, you can use flashbacks.</p>
```

Correct

```
<title>Flashbacks</title>
<shortdesc>Use flashbacks to preserve the state of a system
so that you can return to the same state repeatedly. A flash-
back captures the entire state of a system at the time that
you create the flashback.</shortdesc>
```

Explain Why Users Should Understand the Concept

For concept topics that discuss advantages, benefits, comparisons, relationships, or decision making, explain why it's important that users understand this information.

Incorrect

```
<title>Disaster recovery plans</title>
<shortdesc>There are many benefits to using this product in
your disaster recovery plan.</shortdesc>
```

Correct

```
<title>Disaster recovery plans</title>
<shortdesc>A typical recovery system, which requires a backup
machine for each machine in your environment is costly. A
disaster recovery plan that uses DRM technology can reduce
recovery time and cost by adding a low-maintenance redundant
system.</shortdesc>
```

Reference Topic Short Descriptions

Reference topics typically describe facts about application programming interfaces (APIs), commands, utilities, tools, components, or other objects. To write effective short descriptions for reference information, answer one or more of the following questions about the object that you're describing:

- What does the object do?
- How does the object work?
- What's the object used for or why is it useful?

Define the Object or Explain What the Object Is For

Reference topics provide brief information about facts or objects. In the short description, define or explain what the reference object (command, option, function, part, or component) is used for.

Incorrect

```
<title>AUTOBREW command</title>
<shortdesc>The AUTOBREW command is one of several automation
commands.</shortdesc>
```

Correct

```
<title>AUTOBREW command</title>
<shortdesc>Use the AUTOBREW command to display and set brew-
ing settings such as time, duration, and amount so that you
can brew coffee or espresso drinks automatically.</shortdesc>
```

Describe Why You Would Use the Object

In some short descriptions, you can describe why the object is important or useful, not just what it's used for.

Incorrect

```
<title>Keyboard shortcuts</title>
<shortdesc>You can use keyboard shortcuts for actions that
you normally do with a mouse.</shortdesc>
```

Correct

```
<title>Keyboard shortcuts</title>
<shortdesc>To avoid repetitive strain injuries, you can use
keyboard shortcuts instead of using the mouse. For example,
you can use the shortcut Ctrl+C to copy text instead of se-
lecting text and clicking an icon with the mouse.</shortdesc>
```

Writing Short Descriptions for Converted Content

If you've recently converted your documentation to DITA, don't automatically move the text from the first paragraph to the <shortdesc> element. Automatically moving the first paragraph to the <shortdesc> element might save you time, but it deflects your attention from the quality of the short description content.

Instead, evaluate the content of the first paragraph and verify that it meets the standards for a well-written short description. If the first paragraph doesn't effectively describe the theme of the topic, rewrite the short description.

Alternatively, you can focus on short description standards before you convert content to DITA. If the first paragraph of your topics meets your short description standards, automatically moving this content to the <shortdesc> element can improve your conversion efficiency.

The <abstract> Element

Use the <abstract> element, which is valid between the topic title and the topic body, when the short description of a topic is unsuitable as preview text. You can also use the <abstract> element to introduce a topic if you must include lists, tables, or other elements that are not allowed in the <shortdesc> element.

In the output, the text in the <abstract> element appears as the first paragraph of the topic. However, the abstract text won't be included in preview text.

Use the <abstract> element in these situations:

- To use more DITA elements in the introduction to the topic than the <shortdesc> element allows.

- To include more than one short description in a topic. You can use conditional processing to use the same topic with different short descriptions.

TIP You should use the <abstract> element sparingly. If you do use the <abstract> element, use it consistently and for well-defined situations.

Using More DITA Elements in the Topic Introduction

You can use the <abstract> element to include DITA elements that aren't allowed in the <shortdesc> element. For example, if you create a reference topic that starts with the text of an error message, you should use the <msgblock> element. However, that element isn't allowed in the <shortdesc> element. Instead, use the <abstract> element to include the message content, as shown in Figure 5.15.

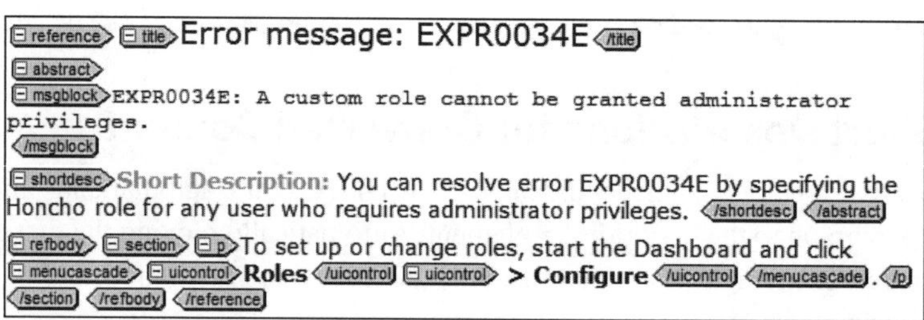

Figure 5.15 A reference topic for an error message that uses both a <shortdesc> and an <abstract> element.

The output of the topic shows both the <abstract> element and <shortdesc> element text as paragraphs, as shown in Figure 5.16.

Error message: EXPR0034E

`EXPR0034E: A custom role cannot be granted administrator privileges.`

You can resolve error EXPR0034E by specifying the Honcho role for any user who requires administrator privileges.

To set up or change roles, start the Dashboard and click **Roles** > **Configure**.

Figure 5.16 The HTML output of a reference topic for an error message that uses both an <abstract> and a <shortdesc> element.

However, the text in the hover text for the related reference link shows only the text from the <shortdesc> element, as shown in Figure 5.17.

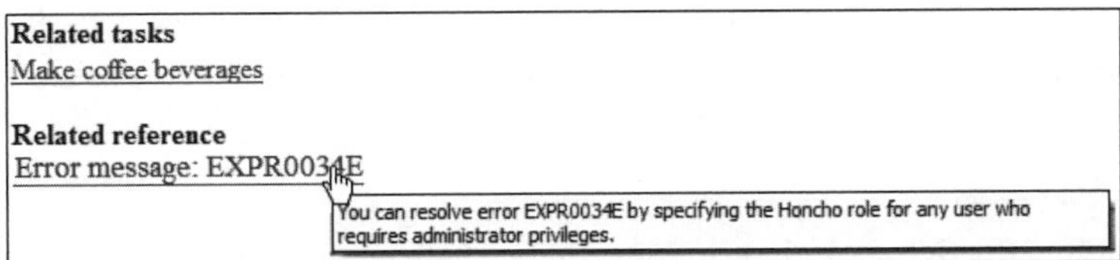

Figure 5.17 The short description in hover text for a reference topic that also includes an <abstract> element.

WATCH OUT Ensure that the abstract doesn't simply repeat or replace the short description. Users don't want lengthy introductions that build up to a main point.

Including Multiple Short Descriptions

Another reason that you might want to use the <abstract> element is to include two or more short descriptions in a single topic. You might want to reuse a topic in different contexts where the short description text must differ. The only location where you can include multiple <shortdesc> elements in a single topic is in the <abstract> element.

Figure 5.18 shows how you can include two <shortdesc> elements by inserting them in the <abstract> element. One short description applies to the Exprezzoh 9000N product and the other short description applies to the Exprezzoh 8500B product.

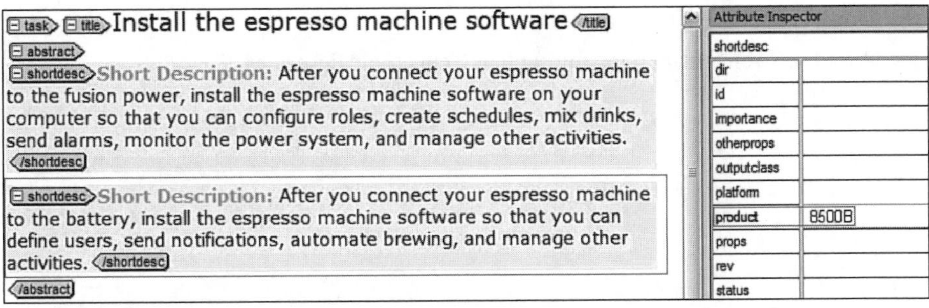

Figure 5.18 Two short descriptions inserted in the <abstract> element.

The text in both <shortdesc> elements is not meant to be read in the same topic. Use conditional processing attributes to specify which <shortdesc> element to include in the output. Each <shortdesc> element has a different value for the product attribute: product = 9000n and product = 8500b. (See Chapter 9, "Conditional Processing," for more about conditional processing attributes.) When you build output of this topic, you'll get a different short description in the topic depending on which conditional value you've specified to include.

A short description is essential for each topic. If you use the <abstract> element, ensure that each topic also contains a <shortdesc> element so that preview text is consistent.

To Wrap Up

Although writing effective short descriptions can be challenging, remember how useful and versatile the <shortdesc> element is. Writing effective short descriptions in your topics can improve the retrievability of your information.

When you write short descriptions, remember that they can be seen in these contexts:

- The first paragraph of a topic
- Hover text for related links
- The text under child topic links
- Search engine results

To write a brief statement that summarizes the topic and works well in different contexts takes a bit of practice and patience, but the benefits are that your users can scan the short description and know exactly what's in the topic.

When you write short descriptions, remember:

- Keep short descriptions short.
- Insert a short description in every topic.
- Don't introduce tables, lists, or figures.
- Use complete, grammatical sentences.
- Don't simply repeat the topic title.
- Avoid self-referential phrasing such as "This topic describes...".

Short Description Examples

Topic Title and Type	Short Description	Comments
Query techniques (Concept)	**Incorrect:** Use one of the following query techniques: **Correct:** SurferDude search collections support a range of query techniques, such as searching for exact phrases, excluding terms, and using wildcard characters.	The incorrect short description introduces a list.
Web browser history (Concept)	**Incorrect:** Your web browser performs many functions other than just helping you visit websites. **Correct:** Web browser history is a list of websites that you visited and are listed by date. Your browser "remembers" the addresses to these sites and helps you access them more quickly.	The incorrect short description doesn't get to the point immediately.
Configuring communication protocols for a local database instance (Task)	**Incorrect:** This task describes how to configure communication protocols for a local database instance by using the Controller Console. Communication protocols on the database server must be configured in order for your database server to accept requests from remote clients. **Correct:** You can configure communication protocols for a local database instance by using the Controller Console to specify that the database server accepts requests from remote clients.	The incorrect short description is unnecessarily long and uses the unnecessary phrase "This task describes...."
Associating search applications with collections (Task)	**Incorrect:** Before you begin the next task, which explains how to use the search application, you must associate it with the collections that it can search. **Correct:** To use the search application, you must associate it with the collections that the application can search.	The incorrect short description refers to another topic and is wordy. Referring to other tasks in this way can prevent a topic from being reused.

Topic Title and Type	Short Description	Comments
Response file keywords (Reference)	**Incorrect:** This topic describes some of the keywords that you can specify when performing a response file installation. You can use the response file to install additional components or products after an initial installation. The following response file keywords are explained with the sample response file. The edited response file must then be copied to your shared network drive or network file system where it can be used by your installation server. **Correct:** To install additional components after the initial installation, you can use a response file. Add the appropriate keywords to the response file and then start the installation program from the command line.	The incorrect short description uses the unnecessary phrase "This topic describes...." and is much too long. The short description also introduces a list.
Hardware and disk space requirements for the Exprezzoh 9000N (Reference)	**Incorrect:** The Exprezzoh 9000N has the following hardware and disk space requirements: **Correct:** Hardware and disk space requirements depend on your operating system, your power source, and the additional accessories that you use for the Exprezzoh 9000N.	The incorrect short description introduces a list and simply repeats the title. Remember that you introduce the list in another section in the topic.

Short Description Checklist

Guideline	Description
Remember that short descriptions are seen in places other than the first paragraph of the topic.	Ensure that short descriptions also make sense in these contexts: • Short descriptions provide hover text for links. • Short descriptions are shown in child topic links. • Short descriptions are sometimes shown in search engine results.
Use full, grammatical sentences.	Follow these guidelines: • Use full sentences with periods. • Follow standard grammar and punctuation rules. • Follow consistent patterns of sentence structure where it makes sense. For example, structure reference topic short descriptions in the same way. In rare cases, you might use incomplete sentences for topics such as API reference information or hardware parts descriptions.
Keep short descriptions short, unique, and relevant.	Follow these guidelines: • Limit short descriptions to one or two sentences and under 35 words when possible. Sometimes using up to 50 words is acceptable. • Don't just repeat the title. • Make each short description unique so that users can discern the subject of the topic solely from the title and short description.
Short descriptions must not reference other content.	Remember that introductions to lists or tables won't make sense when you read the short description text under child links or in hover text for links. Follow these guidelines: • Don't introduce lists, tables, or graphics. • Don't end a short description with a colon. • Don't use self-referential language or make the topic the subject of the short description. • Don't use relational language, such as above or below. • Don't include URLs in the short description.

Guideline	Description
Write short descriptions that are effective for tasks.	For task topics, answer these questions: • What does the task help users accomplish? • What are the benefits of doing this task? • When would users perform the task? • What's involved in the task? • Why do users need to do this task? • When would users perform the task? • How does the task fit with other related tasks?
Write short descriptions that are effective for concepts.	For concept topics, answer these questions: • What is this object, concept, or idea? • Why should users care about this object, concept, or idea?
Write short descriptions that are effective for reference information.	For reference topics, answer these questions: • What does the object do? • How does the object work? • What is the object used for? • Why is this information useful?

Part II

Architecting Content

If you've survived basic training—topic-based writing and short descriptions—and you're now ready for Part II, you must be hungry for more. Excellent! DITA has even more to offer than just the basic framework for topics.

Your next challenge is to architect your information: You need to assemble your topics into information that is retrievable, organized, and reusable. To get there, you'll need to learn about the DITA maps, linking, metadata, and other fun topics.

The concepts in Chapters 6–10 can be a bit daunting, but taking advantage of these powerful DITA features can pay off in the future. For example, you can write fewer topics by reusing what you've already written. You can reduce the maintenance of checking links by using the automatic linking in DITA and by collecting your related links in a relationship table.

You don't need to drink in all these chapters in one sitting. Get comfortable with DITA maps and linking before you try to gulp down metadata and conditional processing.

So grab some espresso and let's get to it.

CHAPTER 6

DITA Maps and Navigation

DITA maps are the glue that binds your topics together, the driver for producing your output, and the information path for your users to follow. Well-structured DITA maps are vital to the usability of online and print documentation because they are the basis for navigating and linking content.

A well-structured DITA map can help users to quickly find the information that they need, provide a task flow that supports the use of your product, and can even provide a few shortcuts for writers. A *DITA map* is an XML file that you use to organize your DITA topics. Use DITA maps to:

- **Include topics in an information set:** Add topics to a DITA map to specify which topics are included in the output.
- **Define information architecture:** Use the DITA map file to define the navigation for a set of topics. Structure topics in a task flow that best meets users' needs.
- **Create relationships between topics:** Use the hierarchy of topics, relationship tables, and other features in the DITA map to establish relationships between topics.

DITA Map Structure

DITA map files have the extension `.ditamap`. The <map> element is the root element of the DITA map file, which contains all the content in the DITA map.

Also, although it's not a required element, each DITA map can include a <title> element. The title helps writers to understand the purpose of the DITA map and is often exposed to users as part of the navigation. As part of your style guidelines for your output, define how the <title>

element in DITA maps is displayed in the output. The content in the <title> element might be displayed in several ways:

- The cover page of PDF output
- The label for a section of topics in the navigation or table of contents
- The title in your web browser, as shown in Figure 6.1

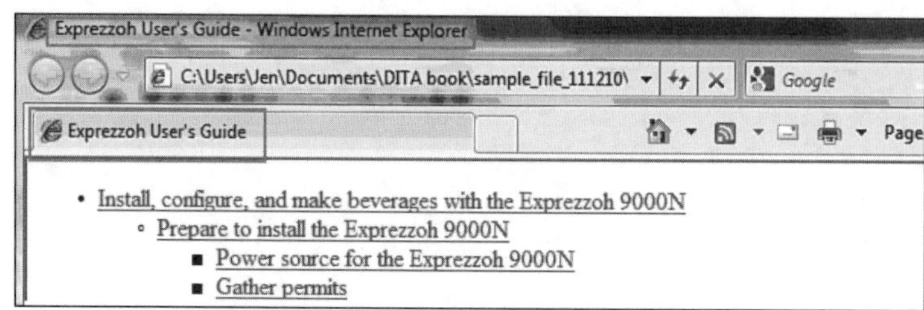

Figure 6.1 DITA map title displayed in a web browser.

Just as you create titling guidelines for DITA topics, create guidelines that explain what text is appropriate in the <title> element of DITA maps to name information sets. For example, should your installation manual be titled "Exprezzoh 9000N Installation Guide" or "Installing the Exprezzoh 9000N"?

Relationships Between Topics

The hierarchy of files in the DITA map defines relationships between two or more topics. For example, when you nest topics inside <topicref> elements, you create family relationships. Topics that contain other topics are called *parent topics*. Nested topics are called *child topics*, and topics nested together at the same level are called *sibling topics*, as shown in Figure 6.2.

Information Organization

Include topics in a DITA map by using a topic reference, or a <topicref> element. The topic reference declares which topics will be included in a particular DITA map. A <topicref> element can reference a topic, DITA map, or even a non-DITA file.

The following example shows a <topicref> element for a task topic with the file name `ext_topic.dita`.

```
<topicref href="ext_topic.dita" type="task">My Task
 Topic</topicref>
```

Figure 6.3 shows how you can nest topics in the DITA map by using the <topicref> element to create a hierarchy of information.

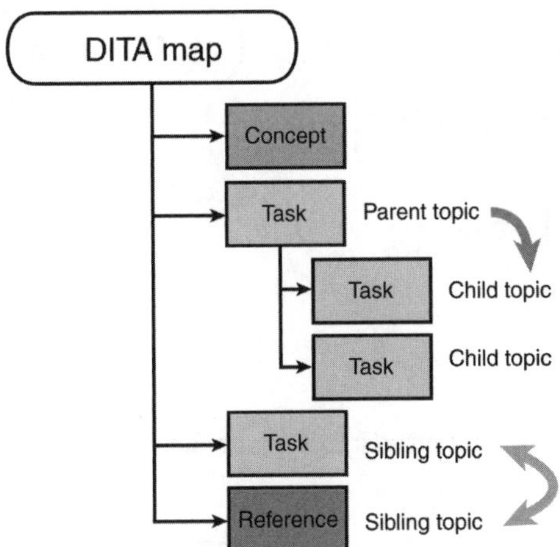

Figure 6.2 Parent and child topics in a DITA map.

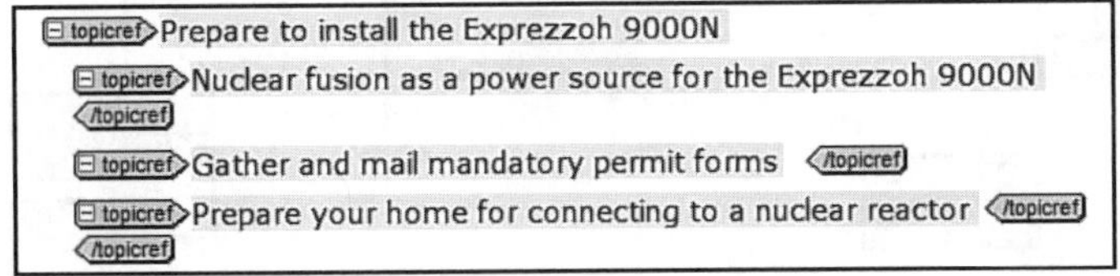

Figure 6.3 Hierarchy of topics in a DITA map.

BEST PRACTICE Avoid nesting topics more than six levels, and even six levels might be pushing it. The more you nest, the harder it is for users to find the topics in the output. Moreover, most output formats get a bit cranky when there are too many levels of nested topics. HTML or PDF viewers can render only so many reasonable font sizes, highlighting styles, and indentation levels to distinguish headings.

You add topics to a DITA map to create a flow of information that best serves your users. For example, you might start organizing your information in a DITA map by including the task topics, as shown in Figure 6.4

Next, you can include concept topics that support the tasks. Connect the supporting concept topics to the task topics to identify related content, as shown in Figure 6.5. This is the start of your web of information.

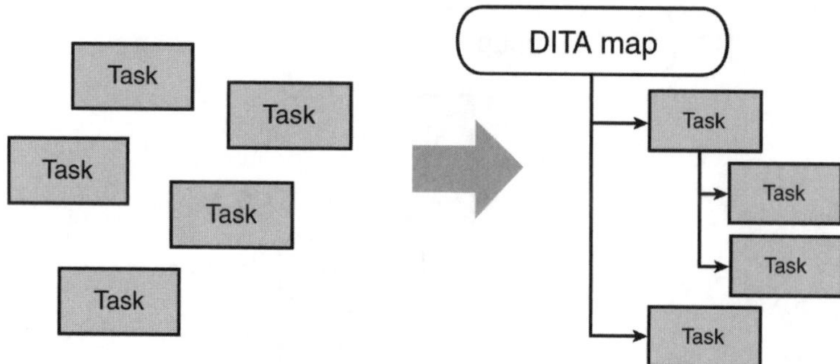

Figure 6.4 Task topics being organized into a sequence.

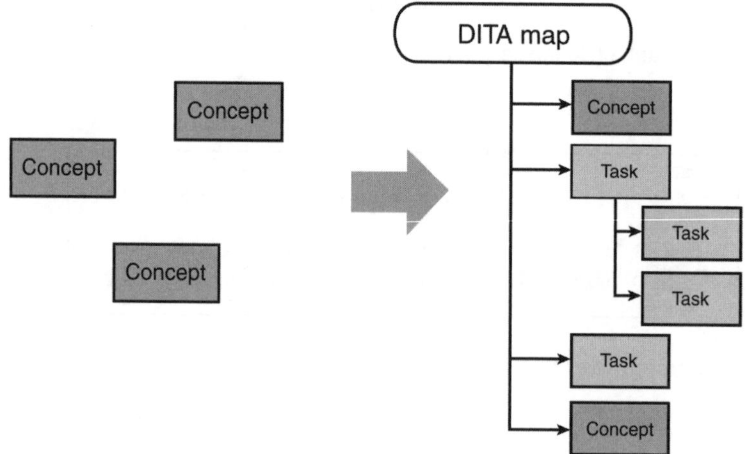

Figure 6.5 Concept topics that support task topics.

Finally, add the supporting reference topics to the DITA map and connect them to the appropriate task and concept topics, as shown in Figure 6.6.

Now that you have your task, concept, and reference topics written and added to the DITA map, you can easily reuse the topics in other information sets by simply creating another DITA map file.

You can combine existing topics in DITA maps to create new information sets. For example, you might include some of the same topics in your online help and PDF information sets, as shown in Figure 6.7.

WATCH OUT Only topics declared in the DITA map file are included in the output. If your topics reference other topics not included as topic references in the DITA map, your output will contain broken references.

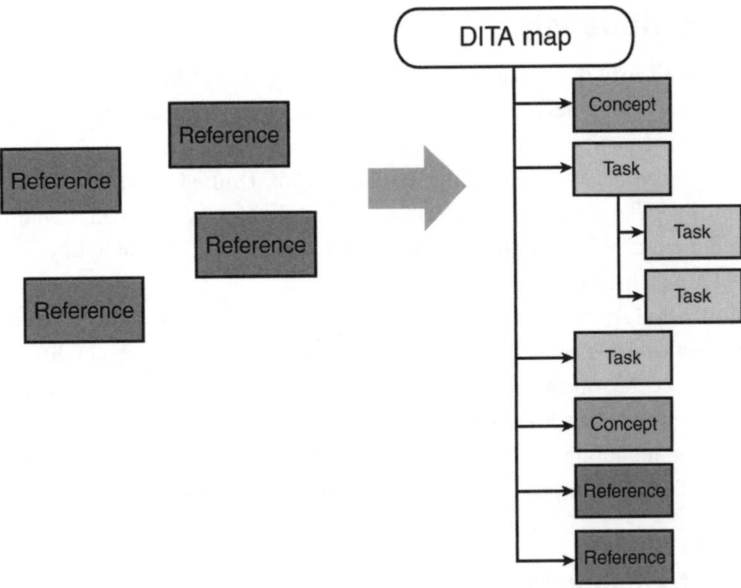

Figure 6.6 Reference topics that support concept and task topics.

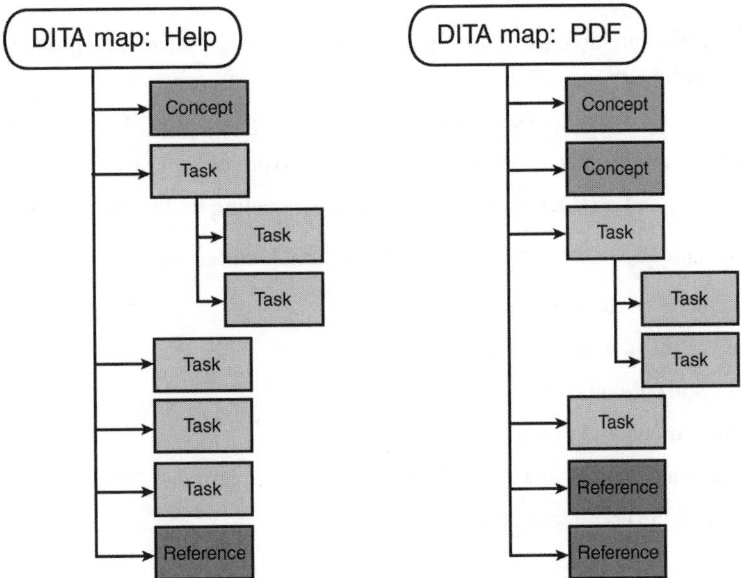

Figure 6.7 Topics that are reused and organized into two information sets.

Information Modeling

No, your tasks won't strut down the fashion show runway. An information model helps you map out what topics go where, what topics are linked, and what sort of linking relationships those topics need.

You can use professional modeling tools, such as Unified Modeling Language (UML) applications or the DITA-aware IBM Information Architecture Workbench, to help you create a well-structured hierarchy for your topics. These tools are visual aids that help you organize your topics into DITA map hierarchies.

The Information Architecture Workbench is an open-source information modeling tool that helps you to structure DITA maps and topics by letting you see the topics in different ways. For example, you can see the topics as a traditional outline that looks like a DITA map, or you can see the topics as a diagram of boxes connected to each other.

Unlike other modeling tools, this Eclipse-based tool is DITA-aware, meaning that the tool can read DITA maps, topics, and specific attributes that you set for those DITA maps or topics.

Benefits of Information Modeling

Information modeling does require resources but is worth the investment if you want to improve the quality of your information. At your company, you might use information modeling to accomplish several goals:

- Information architects can use the modeling tool to organize large groups of topics into information sets.
- Writers can use the modeling tool to review the organization of topics in a DITA map.
- Usability engineers can use the modeling tool to create user task flows and diagrams.
- Engineers and product managers can use the modeling tool to create a use case model.

More specifically, writers can use the Information Architecture Workbench to:

- Structure links between topics
- Analyze the task flow of scenarios.
- Do organizational edits
- Plan documentation projects
- Gather stakeholder agreement about how the information will be organized
- Visualize the user experience

Building Information Models

The Information Architecture Workbench includes a good tutorial that can help you start modeling content, but here's the short version:

1. **Create a modeling project:** To model a set of information, you must create a project. The models and DITA maps that you create in the Information Architecture Workbench are contained in Eclipse project files.
2. **Create a DITA map:** You can create a new DITA map or open an existing DITA map that is stored in a project. You can also import a DITA map file from your file system into a project.
3. **Add topics to a DITA map:** To create a model of your information, add new topics to a DITA map. If you import a DITA map file from your file system, the topics included in the DITA map are displayed.
4. **Reorganize content:** The Information Architecture Workbench is a powerful tool with many options for organizing new content or reorganizing existing topics and DITA maps. After you add topics to a DITA map, you can reorganize the topics and add links.
5. **Export a DITA map:** You can export your modeled content from the DITA map that you created in the Information Architecture Workbench to a standard or custom file type that you can work with in an XML editor.

Figure 6.8 shows how you can view your topic hierarchy and task flow by using a modeling tool.

TIP The Information Architecture Workbench is a free tool. Download it from IBM and try it out: http://www.alphaworks.ibm.com/tech/taskmodeler.

Bookmaps

A special type of DITA map is a bookmap. You can use a bookmap to create PDF output in book-style format that includes chapters, front matter, and back matter. The main components of a bookmap file are the book title, front matter, chapters, appendixes, back matter, and book metadata, which includes information such as the book author, copyright date, and book number. For more information about bookmap metadata, see Chapter 8, "Metadata."

Figure 6.9 shows how a bookmap resembles the structure of a traditional book.

But wait—we know what you're thinking: A typical DITA map looks nearly the same as a bookmap. So why bother using a bookmap? The bookmap is specifically designed for traditional book-like information. You can create a typical book structure in a bookmap DITA map by referencing a DITA map or topic in a <part>, <chapter>, or <appendix> element for each book section.

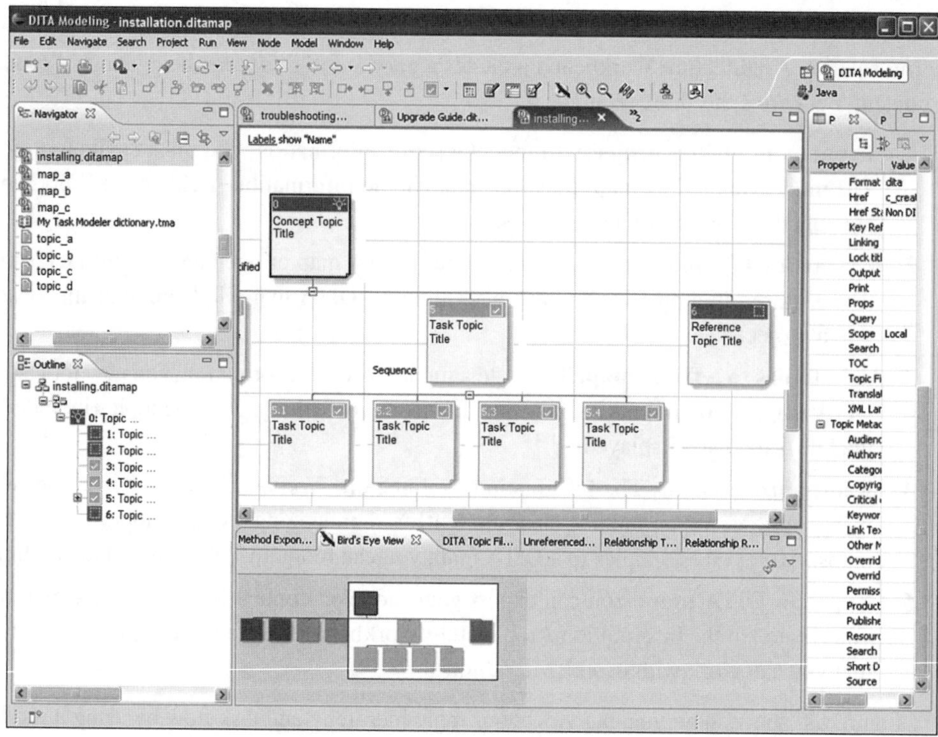

Figure 6.8 A sample DITA map modeled in the Information Architecture Workbench.

If your output is in a book format, you might want to use the bookmap to take advantage of the existing template. However, if you produce multiple output formats, such as HTML for a website or PDF for a book, from the same content or want to reduce the maintenance effort for your writers, consider using a regular DITA map for all output formats.

To set up a low-maintenance book structure by using a regular DITA map, include the necessary book metadata by using the elements and attributes in the <topicmeta> element. Then, add standard information such as copyright holder and year to the output by including the information in your XSL transform, as shown in Figure 6.10.

Submaps

Generally, a single DITA map represents one main user goal. For example, all your installation topics would go in an installation DITA map. However, including all topics for an entire book or help system in one map might make one long DITA map.

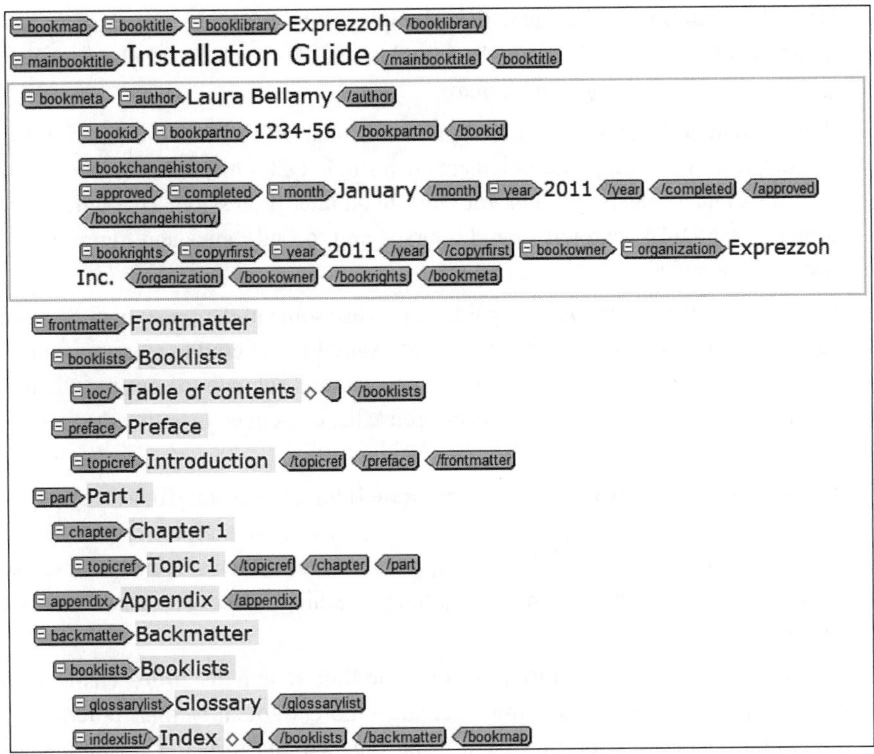

Figure 6.9 A bookmap file in the XMetaL authoring tool.

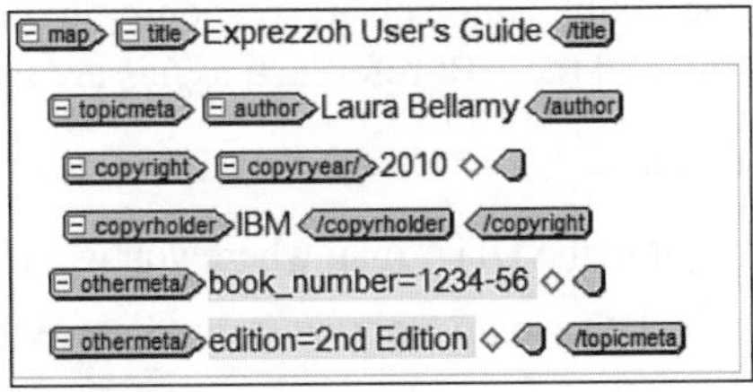

Figure 6.10 Metadata information in a bookmap file.

Instead of creating a gargantuan DITA map for your product, service, or technology, nest smaller DITA maps inside other DITA maps to create submaps. Breaking up your content into submaps helps you to manage your content. Use submaps to:

- **Organize content by chapters:** For example, each chapter in your installation guide might be a DITA map. To create the installation guide, include the chapter submaps in the main installation guide DITA map.
- **Better manage large information sets:** Imagine a help system that has 500 topics. If you include all 500 <topicref> elements in a single DITA map file, the next person who owns that DITA map might look for work in another profession. That unfortunate new owner of the DITA map will spend hours trying to find topics and figure out how the content is organized.
- **Reuse sets of topics:** If two user guides can reuse some of the same topics, you can create one submap for one set of topics. For example, you could create one submap for security topics for a software product. Then, add that submap to your installation guide and administration guide. Then, you can reuse the content by inserting these submaps in other DITA maps.
- **Support peer writers working on the same information set:** To best support writers working on the same groups of topics, organize your content into separate DITA maps so that different writers can work on separate content at the same time. Otherwise, you have to resort to a staring contest or thumb wrestling to see who gets to work on the DITA map first.
- **Segregate frequently updated content so that it can be more easily updated:** Some topics in your library might change more frequently than others. To avoid accessing content that doesn't need to be touched and to better position frequently updated content for quick updates, segregate the content into separate DITA maps. For example, in some software products, the administrative user tasks might change with every release. However, the API reference information remains stable over several releases. Reduce the maintenance effort by including the administrative task topics are in one DITA map, and the API reference topics are in another DITA map.

To add a submap to a DITA map:

1. Insert a <topicref> element in the DITA map where you want to nest the submap.
2. Set the href attribute to the DITA map that you want to reuse.
3. Set the format attribute to "ditamap."

```
<topicref format="ditamap" href="config.ditamap">Configuring the
Exprezzoh 9000N</topicref>
```

DITA Map Ownership

Adding topics to a DITA map is relatively easy, but deciding who owns the organization and how to organize those topics to best help users find the information that they need can be difficult. An information architect can be the best person to own the navigation of your information.

One responsibility of the information architect role might be to create DITA maps with empty *placeholder* topics. Writers are then assigned one or more of those topics and add the correct content to the topics.

An alternative is to identify members of your writing team who are primarily responsible for producing topics and identify those who will consume those topics. Content producers write topics. Content consumers add these topics to DITA maps that support particular products, solutions, or strategies.

Reference Non-DITA Content

You can reference non-DITA content directly from the DITA map by using a <topicref> element. For example, you might want to include a link to a PDF or external website directly from your information.

Figure 6.11 shows a <topicref> element that uses the format, href, navtitle, and scope attributes to reference a PDF file.

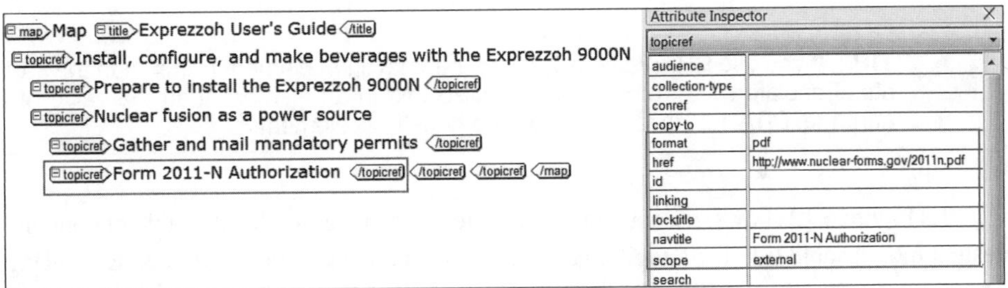

Figure 6.11 A <topicref> element that references non-DITA files.

For more information about linking to non-DITA content and the scope attribute, see Chapter 7, "Linking."

Include Relationship Tables in DITA Maps

You can insert a relationship table in any DITA map. Relationship tables define linking for topics inside or outside the DITA map. For more information about relationship tables, see Chapter 7, "Linking."

Override Topic Titles and Short Descriptions

You can further control and modify the navigation of your information by specifying a different title or short description to display in link previews of topics. These features don't affect the content in the topic, but they help you to customize the navigation.

Navigation Titles

Each topic reference in a DITA map includes a navigation title in the navtitle attribute of the <topicref> element. By default, when you insert a topic into your DITA map, the navtitle attribute is automatically populated with the topic title from the <title> element of the DITA topic. Most of the time, you don't need to change the title.

However, if you want a different title for the topic displayed in the navigation, you can customize the text displayed in the navigation by using the navtitle attribute and lock it by setting the locktitle attribute to "yes."

Don't worry: Using the navtitle attribute doesn't change the title of the topic when that topic is displayed in the output or in the DITA topic itself.

The following code shows how you'd enter a navigation title for a topic in the <topicref> element. Although the <title> element in the topic says "Make coffee, tea, soda, smoothies, juices, and alcoholic beverages," the navigation title provides an alternative title for the navigation:

```
<topicref href="ext__make_beverages.dita" navtitle="Make
 assorted beverages" locktitle="yes">
```

TIP If you use DITA Version 1.2, use the <navtitle> element rather than the attribute. Translators can more easily localize text in an element than the value of an attribute. DITA 1.1 doesn't support the <navtitle> element.

As Figure 6.12 shows, use navigation titles to shorten the title for the table of contents or to create a more usable title for a topic that is reused or duplicated. For example, use a navigation title in a DITA map to provide a navigation title for topics that use the copy-to attribute. See the section "Reuse Strategies" in Chapter 10, "Content Reuse," for more information about the copy-to attribute.

Short Descriptions

You can customize the short description text for a <topicref> element. Adding a short description in the DITA map doesn't override the short description text that is in the topic.

Figure 6.13 shows how you can include a short description in the DITA map by using the <topicmeta> element of the topic reference.

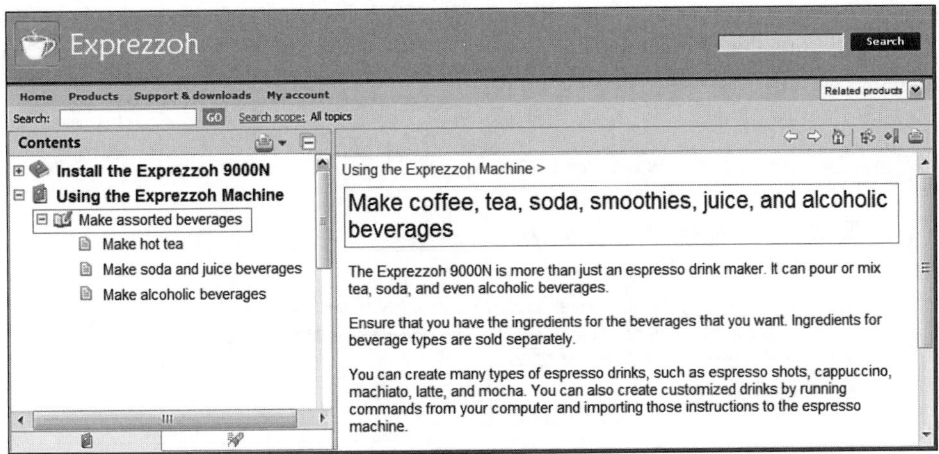

Figure 6.12 A navtitle that provides an alternative title in the navigation tree of a help system.

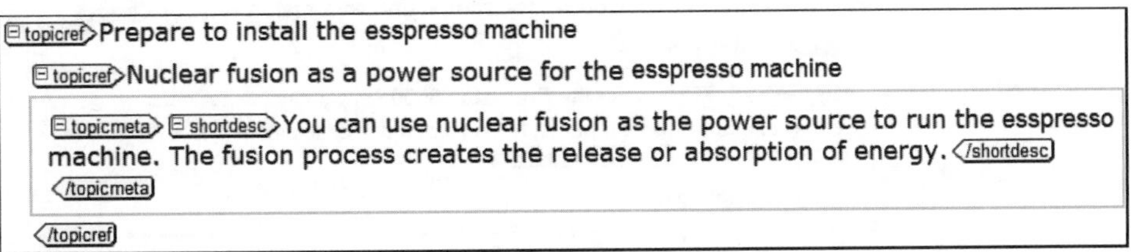

Figure 6.13 Customized short description in the DITA map.

You might want to provide alternative text for link previews in these situations:

- You reuse a topic and want to display a different link preview.
- You include an external link to a non-DITA target.

Provide Unique Short Descriptions for Reused Topics

For topics reused in multiple information sets, you can provide a unique short description for the topic by using the <shortdesc> element in the DITA map as shown in Figure 6.14.

For example, you might reuse the task topic "Gather permits," which describes how to gather required permits in two separate sets of information. If both information sets are available to users at the same time in an HTML system, users might receive two search results that seem to point to the same "Gather permits" topic. But, if you provide alternate short descriptions, your users can determine that the first "Gather permits" topic is for the Exprezzoh 9000N Installation

Guide, whereas the second one is for the Exprezzoh 8500B Installation Guide as shown in Figure 6.15. By providing this alternate preview text for the reused topic the user can correctly choose the "Gather permits" topic in right set of topics.

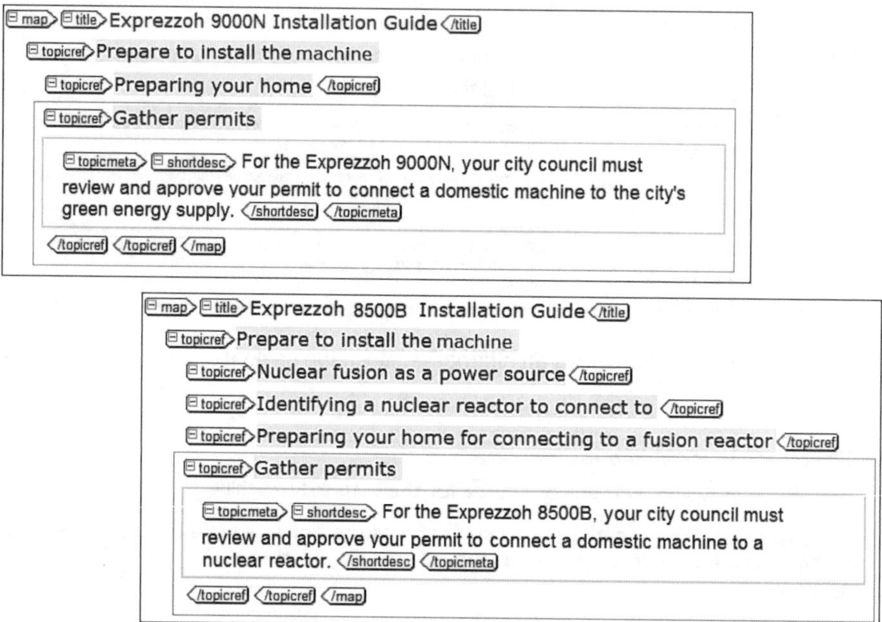

Figure 6.14 A unique short description for reused topics.

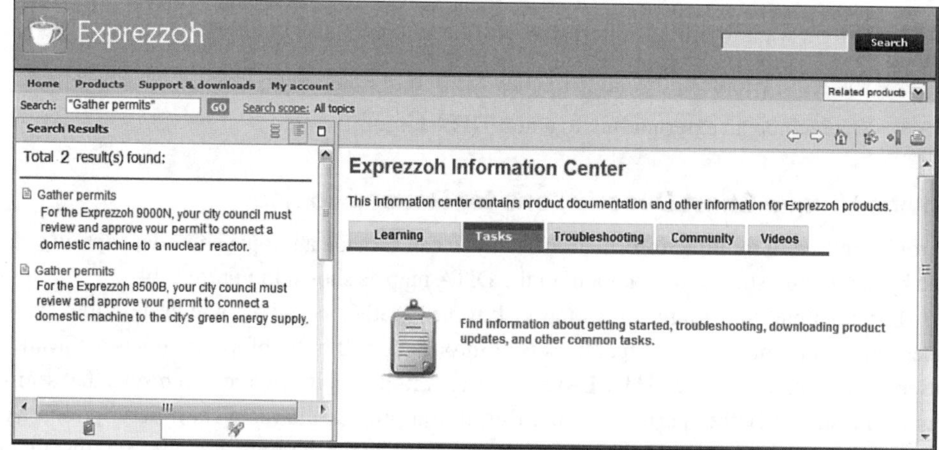

Figure 6.15 An HTML system that displays unique short description for duplicate topics.

Provide Short Descriptions for Links to Non-DITA Content

You might want to link to non-DITA content, such as PDF, HTML, or other file types. When you include these links in the DITA map <topicref> elements, you can also include a <shortdesc> element. This way, even links to non-DITA content can include link previews that are displayed as hover text.

Figure 6.16 shows a <topicref> element that links to an external website. The URL to the website is specified in the href attribute of the <topicref> element.

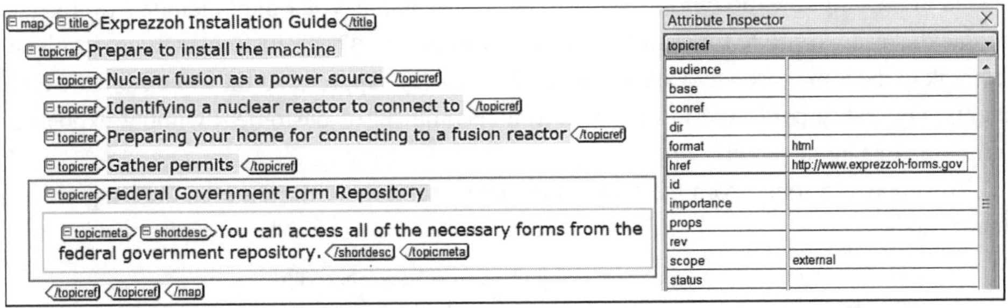

Figure 6.16 A <topicref> element that links to an external website.

When you provide short descriptions for links outside of your information set, follow these guidelines:

- Be specific enough to let users know whether the link provides the information that they're looking for.
- Help users find the right information if the target page is long and the information that they need is buried somewhere on that target page.
- Provide enough information to help users know where the link is going in case the link breaks. If the link is broken, users might search from a commercial search engine.
- Keep short descriptions under 35 words when possible. In some browsers, the hover text is truncated or quickly disappears.
- Use consistent phrasing for short descriptions that go outside your information set. For example, use "See the *some_target_site_or_page* for information about *this_subject*."

The following example shows incorrect and correct short descriptions in DITA maps for external links:

Incorrect

```
<title>Exprezzoh Drinks FAQs</title>
<shortdesc>Learn more about brewing espresso.</shortdesc>
```

Correct

```
<title>Exprezzoh Drinks FAQs</title>
<shortdesc>See the Espresso Drinks website for quick answers
to common questions about brewing techniques.</shortdesc>
```

Suppressing Topics from the Table of Contents

Sometimes, you don't want a topic shown in the table of contents, but you do want the topic to be included in the output with your information set. For example, you might include a troubleshooting topic that is appropriate for a small set of your customers. These customers might be given the direct link to the topic when they encounter a problem. By suppressing the topic from the table of contents, you can provide the information to a specific set of customers without exposing the information to a larger set of users.

To suppress a topic from the table of contents, set the toc attribute on the <topicref> element to "no."

```
<topicref href="ext_topic.dita" toc="no">My Task
 Topic</topicref>
```

WATCH OUT The value of the `toc` attribute is inherited by child topics. If you apply this property to a parent topic, you'll also suppress the child topics from appearing in the table of contents. On each child topic that you want to see in the table of contents, you must set the toc attribute to "yes" to restore the default behavior of those child topics.

Suppressing Content from PDF Output

You also might not want a topic to be shown in PDF output, but you do want the topic to appear in HTML output. Through the wonders of DITA, you can suppress topics from appearing in one output format but allow them to appear in another.

You'll find this feature handy especially if you use the same DITA map for PDF and HTML output. Rather than using conditional processing attributes to suppress topics from the PDF output, use the print attribute.

To suppress content from PDF output, set the print attribute on the <topicref> element to "no."

```
<topicref href="ext_topic.dita" print="no">My Task
 Topic</topicref>
```

WATCH OUT Once again, inheritance can catch you by surprise. The value of the print attribute is inherited by child topics. If you apply this property to a parent topic, you'll also prevent the child topics from appearing in the PDF output. On each child topic, you must set the print attribute to "yes" to restore the default behavior of those child topics.

Suppressing Content from HTML Output

For the same reason that you might want to suppress content from PDF output, you might also want to suppress some content from HTML output. Set the print attribute to "printonly" to prevent the topic from appearing in HTML output.

For example, in case you're not ready to part with book-like content that glues chapters together, you can leave that content in PDF output but remove it from the HTML output. In this case, you can set the print attribute on the <topicref> element to "printonly."

```
<topicref href="ext_topic.dita" print="printonly">My Task
 Topic</topicref>
```

WATCH OUT As of the publish date of this book, the printonly feature was not working as designed in DITA Open Toolkit publishing.

To Wrap Up

You'll likely find that working with DITA maps is fun and easy. OK, "fun" might not be the right description, but DITA maps help you organize topics, insert new topics where you want them, and see the structure of your information set. Also, many DITA authoring tools provide several ways to view your DITA map and architecture, such as an outline view or a hierarchical view.

Remember that you can also include a different short description in the <topicmeta> element of a topic reference if you want to:

- Customize the link preview for a topic
- Provide a link preview for an external link

Before you have too much fun with DITA maps, be sure that you first architect your information. Find a talented information architect or use modeling tools to figure out the structure of your content. We'll keep harping on this note: Plan before you implement. Model your content before you convert content to DITA or before you create a new information set.

And whether you're converting content to DITA or starting on a new document, you need to decide what types of DITA maps that you want to create: one DITA map from which you create multiple output formats or separate DITA maps for each output format. If you decide to use one DITA map with multiple output formats, you also need to decide whether to suppress some topics from appearing altogether, or whether you don't want them to appear in the table of contents.

Navigation and DITA Maps Checklist

Guideline or Decision Point	Description
Model your architecture before you start writing.	Use information modeling tools or just pen and paper to decide which tasks to cover and in what order they should be presented to the user.
Decide who should own each DITA map.	Determine the owner of the DITA maps: Should each writer own a DITA map or should the architect or primary writer own DITA maps?
Decide whether to create a bookmap for PDF output.	Decide whether to have a separate bookmap for PDF output and a regular DITA map for HTML output.
Decide how to structure each DITA map.	Create one DITA map per user goal. Insert submaps as necessary.
Change navigation titles, short descriptions, or link text as needed.	Update titles when you want them to be shorter in the navigation. Provide alternative preview text of duplicated topics or preview text for external resources that you are linking to.

CHAPTER 7

Linking

Well-written topics by themselves aren't enough to help users accomplish their goals. Writing good topics is only the first step to creating useful technical information. The next step is to help users find that information and to help them get from one topic to another.

By implementing a well-designed linking strategy, you create an effective web of information and improve the retrievability and navigation of your content.

DITA has powerful linking features that you can use to create links:

- **Hierarchical links** reveal the navigation of your information through nested topics.
- **Inline links** connect users to related information; these links are inserted in the text of the topic.
- **Related links** connect users to related information; these links are typically inserted in relationship tables.
- **Collection-type links** create links among a group of nested topics; collection types are set in the DITA map.

Hierarchical Links

When you nest topics in the DITA map, which means inserting <topicref> elements inside other <topicref> elements, you create a topic hierarchy: parent, child, and sibling topics. The topics in this hierarchy are automatically linked to each other creating *hierarchical links*.

By default, DITA creates links in HTML output from a parent topic to its child topic. In the HTML output, these links are displayed as parent topic links. These hierarchical links help users to understand the organization of your information.

Figure 7.1 shows how nested topics in a DITA map create a hierarchy.

topicref Install the espresso machine Parent topic
topicref Set up the hardware /topicref Child topic
topicref Connect the espresso machine to a nuclear reactor /topicref Child topic
topicref Install the espresso machine software /topicref Child topic
/topicref

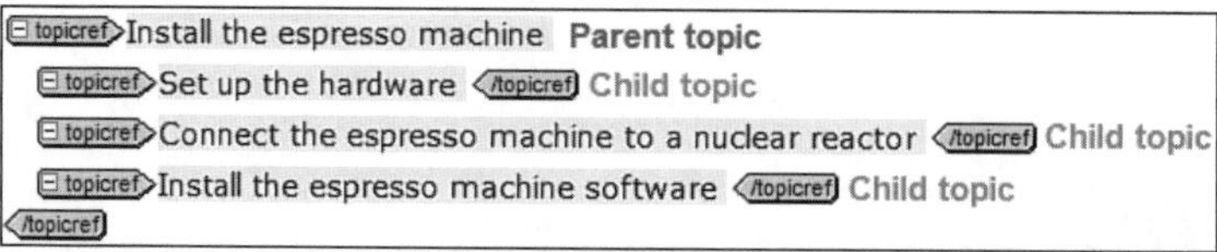

Figure 7.1 Nested <topicref> elements in a DITA, map that create a parent and child relationship.

When you transform the DITA map to HTML output, you see the hierarchy of links in the table of contents or navigation tree, as shown in Figure 7.2.

Install the espresso machine
Set up the hardware
Connect the espresso machine to a nuclear reactor
Install the espresso machine software

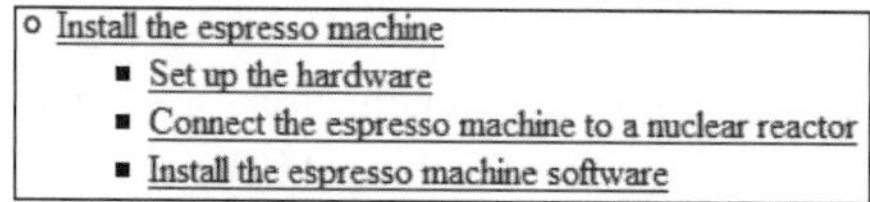

Figure 7.2 Topics in a linked hierarchy in HTML output.

In addition, if you view the topic "Install the espresso machine," you see a link to its parent topic "Install, configure, and make beverages with the Exprezzoh 9000N," inserted at the bottom of the topic as shown in Figure 7.3.

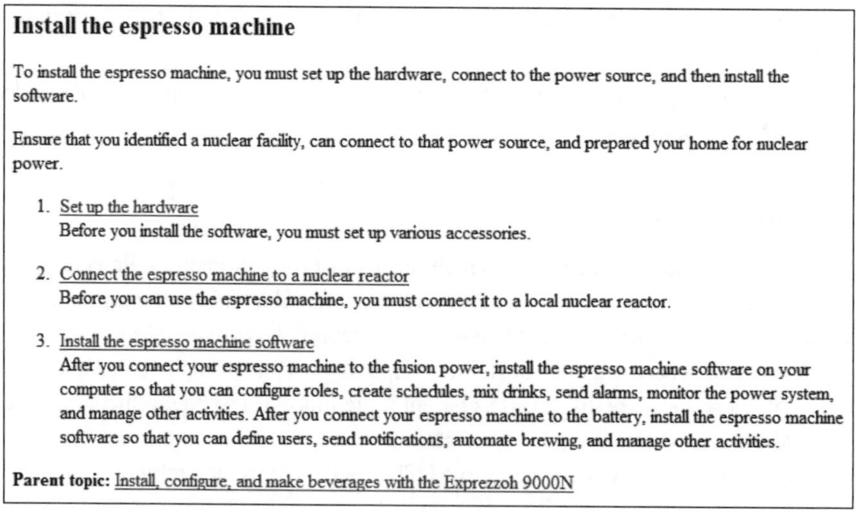

Install the espresso machine
To install the espresso machine, you must set up the hardware, connect to the power source, and then install the software.
Ensure that you identified a nuclear facility, can connect to that power source, and prepared your home for nuclear power.
1. Set up the hardware
Before you install the software, you must set up various accessories.
2. Connect the espresso machine to a nuclear reactor
Before you can use the espresso machine, you must connect it to a local nuclear reactor.
3. Install the espresso machine software
After you connect your espresso machine to the fusion power, install the espresso machine software on your computer so that you can configure roles, create schedules, mix drinks, send alarms, monitor the power system, and manage other activities. After you connect your espresso machine to the battery, install the espresso machine software so that you can define users, send notifications, automate brewing, and manage other activities.
Parent topic: Install, configure, and make beverages with the Exprezzoh 9000N

Figure 7.3 Parent topic links connect the child to the parent topic.

TIP If you don't like the label "Parent topic," you can change it by modifying the style sheet that you use to transform your content to HTML output.

As an alternative to parent topic links, consider using breadcrumbs to show the hierarchy of topics in your HTML output. Figure 7.4 shows how *breadcrumbs* create a trail of links to help users to navigate to topics in the same hierarchy without having to scroll to the bottom of a topic to discover the parent topic link.

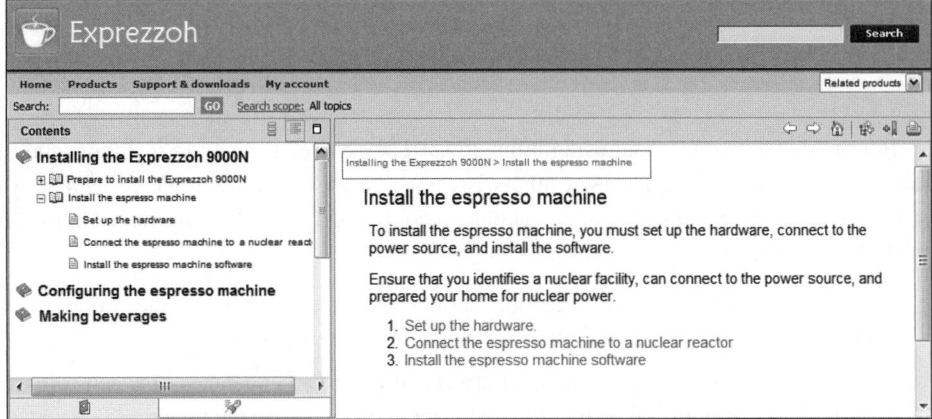

Figure 7.4 Breadcrumb links that help users understand the hierarchy and progression of information.

Inline Links

When you first think of links, you probably think about hyperlinks that appear inline with the text in the middle of topics. *Inline links* are cross-references that you insert in the topic rather than create the link by using a DITA map or a relationship table.

Use the <xref> element to create inline links. Include information about the link target in the href attribute.

```
<xref href="target_topic.dita">
```

Figure 7.5 shows an inline link that uses the <xref> element to create a cross-reference from a DITA task topic to another task topic.

Figure 7.6 shows that in the output of this topic, the cross-reference is a hyperlink that directs users to the referenced file or location.

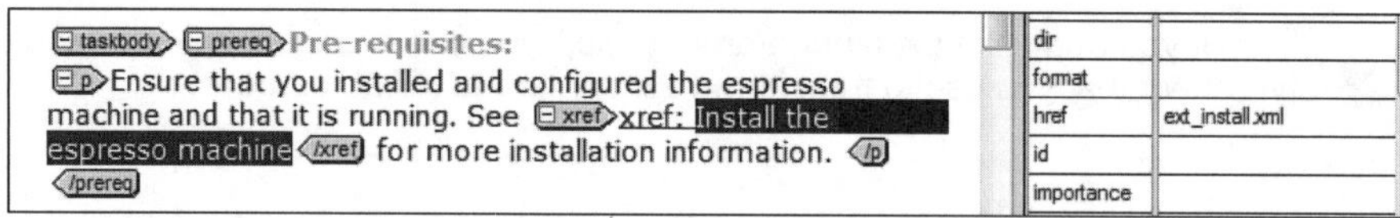

Figure 7.5 An inline link to a topic called "Install the espresso machine."

Ensure that you installed and configured the espresso machine and that it is running. See Install the espresso machine for more installation information.

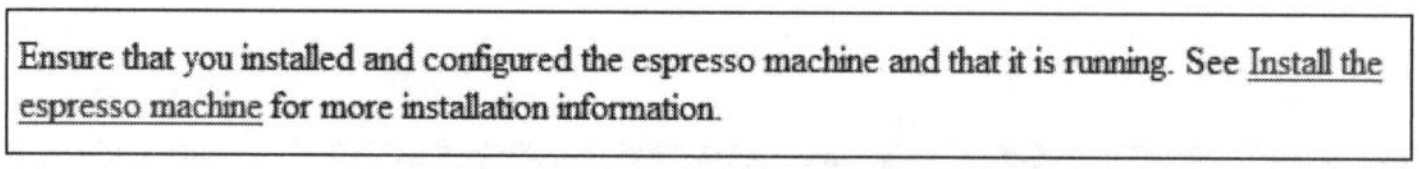

Figure 7.6 The HTML output of an inline link to a topic called "Install the espresso machine."

You should be careful about using inline links: They can be difficult to maintain, and they can distract users. Every link, inline or otherwise, must be useful and relevant. Otherwise, your carefully constructed web of information becomes a confusing maze. Limit the use of inline links for the following reasons:

- **Inline links are disruptive:** They break the flow of reading because they direct users to other topics before users have finished reading the current topic.
- **Inline links create dependencies between topics:** Dependencies between topics are difficult to maintain and restrict to what extent you can reuse the topics.
- **Inline links can complicate navigation:** Numerous inline links often indicate a problem with the topic organization. If you find yourself adding "See topic X for more information" in every paragraph, reevaluate the organization and the purpose of the topic. You might need to rewrite it.
- **Inline links weight content:** Links tell users what's important, similar to using labels on paragraphs such as "Restriction" or "Important." Overusing links, just like overusing labels, makes everything important and therefore nothing is important.

When you do need to use inline links, follow these guidelines:

- Link to prerequisite and postrequisite information.
- Avoid inline links to tables and figures in a topic.
- Create inline links to repeated steps.
- Create inline links to high-level tasks.

Link to Prerequisite and Postrequisite Information

You can point users to prerequisite information by using an inline link.

To create an inline link to a prerequisite topic, in the <prereq> element, insert an <xref> element that references the target topic, as shown in Figure 7.7.

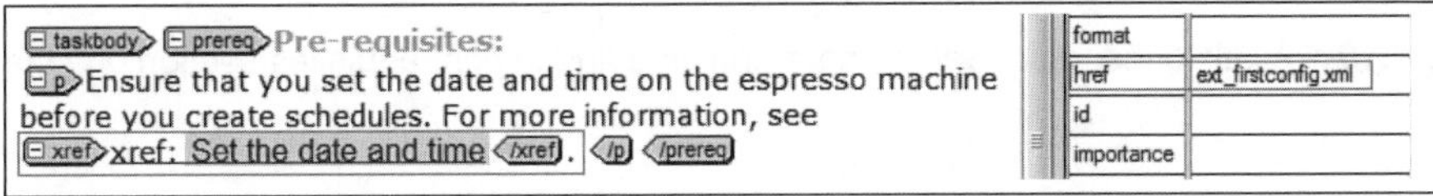

Figure 7.7 An <xref> element inserted in the <prereq> element of a topic.

In HTML output, users can click the link if they need to see the prerequisite information, as shown in Figure 7.8.

Create schedules

If you want your espresso or other drinks to be brewed or poured at specific times, you can set up schedules.

Ensure that you set the date and time on the espresso machine before you create schedules. For more information, see Set the date and time.

Figure 7.8 HTML output of a link inserted in the <prereq> element.

TIP Set the importance attribute on a <topicref> element to “required” to pull a generated link into the <prereq> element of a task topic when you build output. The target topic that you want to link to must be nested in the same topic hierarchy as the source topic that will have the link. See “Links Created with the Important Attribute” later in this chapter for more information.

Similar to inline links for prerequisite information, you can insert an <xref> element in the postrequisite element of a task topic, as shown in Figure 7.9.

postreq Post-requisites:
p After you install the espresso machine software, you must set the time and the date on the espresso machine. For more information, see
xref xref: Set the date and time /xref . /p /postreq /taskbody /task

format	
href	ext_firstconfig.xml
id	
importance	

Figure 7.9 An <xref> element inserted in the <postreq> element of a task topic.

In HTML output, users can click the link to see the postrequisite information, as shown in Figure 7.10.

After you install the espresso machine software, you must set the time and the date on the espresso machine. For more information, see Set the date and time.

Figure 7.10 A link to the topic "Set the date and time" in the postrequisites section of a task topic.

WATCH OUT Be sure not to create duplicate links. Don't create an inline link if the linking relationship between the source and target topics is already defined by hierarchical links or in a relationship table. Duplicate links aren't caught by the DITA Open Toolkit when you build output. Check your output to verify that you didn't inadvertently create the same link twice.

Avoid Inline Links to Tables and Figures in a Topic

Inserting links from a paragraph to a table or figure inside the same topic is typically unnecessary if you're writing short, self-contained topics.

Also, these links can become more difficult to maintain when you update topics. For example, if you delete a table or move it to another topic, you'll also need to update or delete the link. Inline links to tables or figures might also make your topic or the information in the topic more difficult to reuse.

For example, avoid creating links such as the reference to a table, shown in Figure 7.11.

To redirect power to household appliances, you must have the appliance name and ID available. The ID is next to the appliance serial number. The Exprezzoh 9000N is preprogrammed to connect to hundreds of commercially available appliances.

Table 1 describes power management commands.

Table 1. Power management commands

Task	Command	Example
Get reports at various intervals to view power consumption.	`exprezzoh report powerLevel time` For the **time** parameter, you can specify to get reports every hour, every 12 hours, every day, every week, or every month: **hour** A report is created every hour. **12 hour** A report is created every 12 hours.	`exprezzoh report powerLevel 12 hour`

Figure 7.11 An unnecessary inline link to a table.

However, if you need to link to a table or figure that appears in another topic, you have the following options:

- Use an <xref> element to create an inline link to the table or figure.
- Use a content reference (conref) to include the table or figure directly in your topic. See Chapter 10, "Reuse Strategies," for more information about content references.

Create Inline Links to Repeated Steps

In task topics, you occasionally need to write statements such as "Repeat Step 8 for each server in your environment." To ensure that you direct the user to the correct step regardless of how many changes you make to the steps of the task, use an inline link.

In Figure 7.12, step 5 contains an inline link to step 4.

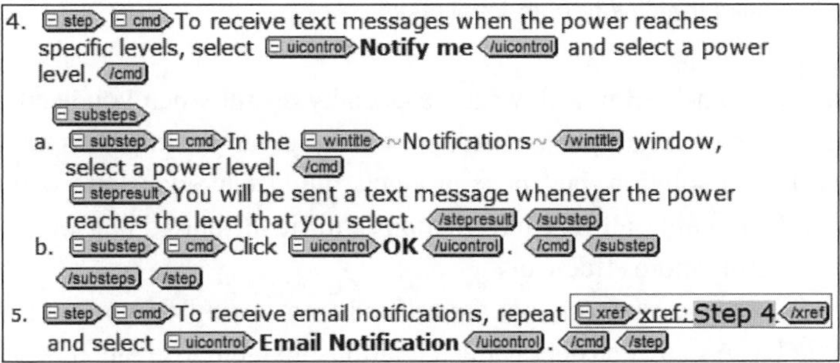

Figure 7.12 An inline <xref> element inserted in step 5 to link to step 4.

By creating an inline link with the <xref> element, you don't need to worry about whether the number for step 4 changes. If a writer later adds or removes steps that change the numbering, the link text changes automatically.

Create Inline Links to High-Level Tasks

To show your users a high-level task flow for more complex products, you can create a roadmap that includes links to each task in the flow. For example, a large database product might have 10 separate task topics for installation, another 15 task topics for postinstallation configuration, and yet another set of tasks for setting up security. These tasks might be spread throughout your DITA map, so you can't take advantage of default linking.

To help users understand the big picture of the task flow, you can create links to each parent task topic. Showing users the overall task flow for the entire product or component helps users understand and plan for the tasks that they need to complete.

Figure 7.13 shows a task flow with inline links for the espresso machine tasks. Each step links to a topic.

> **Install, set up, and make beverages with the Exprezzoh 9000N**
>
> The Exprezzoh 9000N does more than just help you make coffee beverages. This scalable, cross-dimensional product can display calorie, caffeine, and other nutrition facts, feed your pets, power most other appliances in your home, play your favorite music, and do many other tasks.
>
> The Exprezzoh 9000N is a powerful household tool because it uses nuclear energy as its power source. Just hook it to a nuclear fusion reactor and you are ready to make great coffee and other beverages.
>
> To set up the Exprezzoh 9000N, do these main tasks:
>
> 1. Connect to a nuclear reactor.
> 2. Meet all software and hardware requirements.
> 3. Install the espresso machine and its software.
> 4. Set the date and time on the machine.
> 5. Create roles for users.
> 6. Monitor the system.
> 7. Make beverages.

Figure 7.13 Inline links that link to high-level tasks.

Inline links for high-level task flows are especially useful when you need to link to documentation that is created by multiple sources. For example, large companies might offer several products that work as a solution, but those products might have disparate writing teams from around the world. Combining all that information is difficult, but adding inline links can help connect your information more effectively.

To create high-level task flows that are created from a single source, such as pulling the links from a single DITA map, consider using a relationship table to create this task flow.

Control How Links Are Displayed

Sometimes, you might not want certain links to be displayed in the output. For example, you might want links to go from an installation topic to an uninstallation topic, but not the other way around. In this case, you can use the linking attribute to create links that go only one way.

To suppress links from being displayed, set the value for the linking attribute on the parent <topicref> element in the DITA map hierarchy or relationship table to one of these values:

- **Sourceonly:** A topic can't be linked to but can link to other topics.
- **Targetonly:** A topic can only be linked to and can't link to other topics.
- **None:** A topic can't be linked to or link to other topics.
- **Normal:** A topic can be linked to and can link to other topics. This is the default behavior when nothing is set on the linking attribute.

WATCH OUT Be aware that the value of the linking attribute is inherited by the child topics. This might mean that you're setting the linking behavior for topics that you don't mean to.

The linking attribute provides you with more granular control over the linking between your topics so that you can refine the information experience for your customers. Table 7.1 explains when you might want to use the linking attribute.

Table 7.1 Use Cases for the Linking Attribute

Use Case	Linking Attribute Value	Description
Provide no link back to the originating topic.	Targetonly	Use this value to specify that a link to the target topic doesn't also have a link back to the originating topic. This value is most often used in reference information.
Hide a topic.	None	Use this value with the toc attribute to hide topics from some users. For example, if you created a topic about installing a software patch for a specific customer, you might want to suppress all linking to and from this topic but provide the exact URL to this topic for the one customer who requested the software patch.
Restore default linking.	Normal	Use this value to override the linking value of a parent topic. This value is most often used to restore normal linking behavior for a child topic that has a parent topic with its linking attribute set to "none."
Hide a topic from a collection.	Sourceonly	Use this value to suppress a topic from automatically generated linking behavior. For example, if you want to nest a concept topic under a task parent topic but want to suppress the concept topic from being included in the collection of child topic links, set the concept topic to "sourceonly."

Related Links

You probably think by now that hierarchical linking and inline linking are all you need to create a well-organized, well-linked web of information. However, DITA has yet another option for linking: related links.

Use a *relationship table* to create links to related topics. For example, if you have a topic for installing a product in one DITA map and a topic for uninstalling the product in a different DITA map, you might want to create a related link between the two topics.

Related links that are built with relationship tables can further improve the usability and retrievability of your information.

Relationship tables also have the following advantages for writing teams:

- **Centralized link management:** Relationship tables are allowed only in DITA map files. Relationship tables make links between topics easy to locate because the linking relationships are defined in one place, not in each topic.
- **Reduced maintenance:** Because the linking relationship between topics is easy to locate, repairing and updating links from a central location is easier than hunting through individual topics.
- **Improved reusability:** By creating links from the DITA map instead of inside the topic, the topic has fewer dependencies to other files. With fewer dependencies, you improve the reuse potential of the topic.

Use the <reltable> element in a DITA map to create a relationship table. Add topic references in the rows of the relationship table for the topics that you want to link to.

TIP In most XML authoring tools, you can drag and drop topic references from a file directory to the relationship table. You can also copy and paste <topicref> elements from the DITA map hierarchy, but be aware that any attributes set on the topic reference in the hierarchy will also be set on the topic reference in the relationship table. For example, if you set the linking attribute to "none" in the <topicref> element in the DITA map and you copy that <topicref> element to the relationship table, the linking attribute remains set to "none." Therefore, the linking will be turned off unless you change the linking attribute value.

Relationship Tables

The key to creating related links is to understand how the table structure defines relationships. Relationship tables define relationships between topics in different cells of the same row.

TIP You can define relationships on table columns. However, understanding how these relationships work can be confusing. We recommend that, initially, you limit yourself to mastering relationships with cells in the same row.

Figure 7.14 shows how in a basic relationship table, the relationship is between the cells in a row of a <reltable> element. However, no relationship is specified between the cells in each column.

The relationship table in Figure 7.14 defines the following linking relationships:

- Topic1 + Topic2 = Related link from Topic1 to Topic2 and from Topic2 to Topic1
- TopicA + TopicB = Related link from TopicA to TopicB and from TopicB to TopicA
- Topic1 + TopicA = No related link
- Topic2 + TopicB = No related link

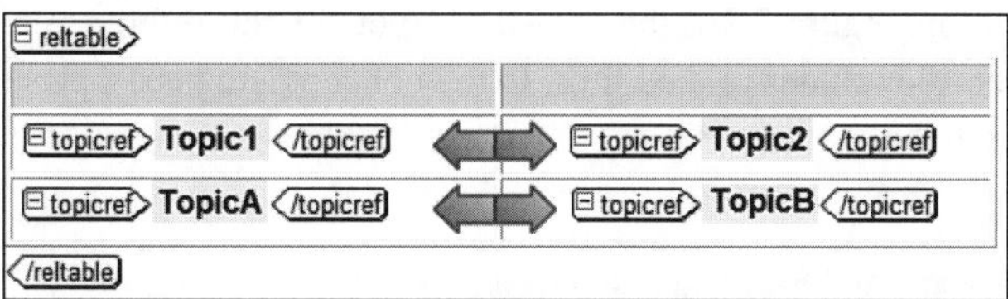

Figure 7.14 Relationships that are defined between cells in the same row.

When you create links in a relationship table and build output of the information set, related links are added to the topics. The related links appear below the body text of the topic and, by default, are labeled according to the topic type as "Related tasks," "Related concepts," and "Related reference."

You can create different types of relationship tables depending on how you want related links to work:

- **Topic-type link tables:** Use topic-type tables to structure related links according to the task, concept, and reference topic types.
- **One-way link table:** Use one-way tables to create related links from one topic to another. The link works only from the source topic to the target topic, but not the other way around.
- **Two-way link tables:** Use two-way tables to create related links between two topics regardless of topic type. Links work in both directions: to and from the source and target topics.
- **Custom tables:** Create your own custom table to create a unique structure that best suits your information.

Topic-Type Relationship Tables

A typical topic-type table contains three columns. Each column represents the standard topic types of task, concept, and reference, as shown in Figure 7.15.

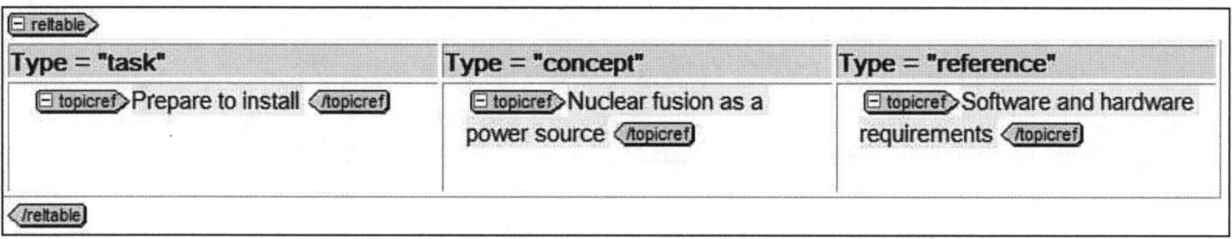

Figure 7.15 A three-column topic-type relationship table in a DITA map.

In the example in Figure 7.16, the HTML output for the reference topic "Software and hardware requirements" includes related links to the concept and task topics that are in the same row of the relationship table.

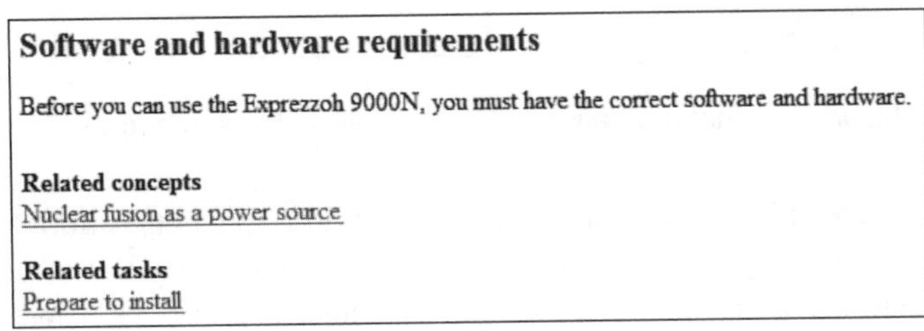

Software and hardware requirements
Before you can use the Exprezzoh 9000N, you must have the correct software and hardware.
Related concepts
Nuclear fusion as a power source
Related tasks
Prepare to install

Figure 7.16 A three-column relationship table that creates related links to related concept and task topics.

Not every topic requires related links to concept, task, and reference topics. You do not need to include a <topicref> element in each column of a row—empty table cells are fine. For example, if you want to create a two-way related link between a concept and task topic, you can leave the reference column in that row empty.

TIP The topic-type, three-column relationship tables can be difficult to understand because the topic type adds complexity. To make relationship tables easier to use and understand, use only two-column relationship tables for one-way linking and for two-way linking.

One-Way Linking Relationship Tables

In a two-column relationship table, you can create one-way links by setting the linking attribute on the first column to "source-only" and the second column to "target-only," as shown in Figure 7.17.

reltable

Linking = "sourceonly"	Linking = "targetonly"
topicref Gather permits /topicref	topicref Identify a nuclear reactor to connect to /topicref
topicref Install the espresso machine /topicref	topicref Set up the hardware /topicref

/reltable

Figure 7.17 A two-column relationship table that uses one-way linking.

In the HTML output, the "Gather permits" topic includes a related link to the topic "Identify a nuclear reactor to connect to," but not the other way around, as shown in Figure 7.18.

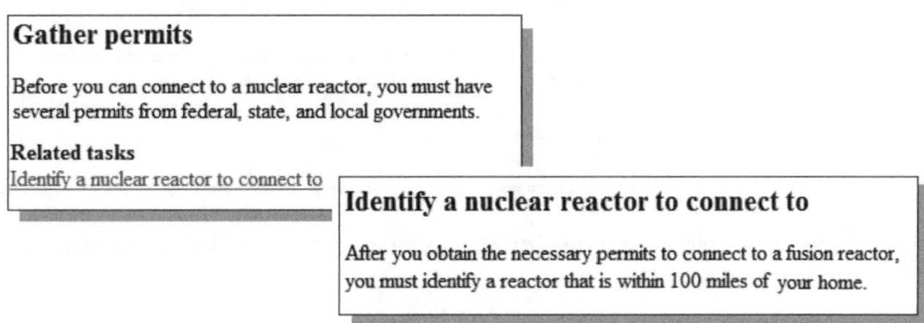

Figure 7.18 Output example from a two-column relationship table that uses one-way linking.

In a simple, two-way linking relationship table, don't worry about putting specific topic types in specific columns. You can insert <topicref> elements for any topic type in either column. The transform to HTML will place your topics under the appropriate heading. For example, all the related task topics will be placed under the default heading "Related tasks" regardless of which columns you've inserted the <topicref> elements.

Two-Way Linking Relationship Tables

In a two-column relationship table for two-way linking, you don't need to set any attributes on the columns for the topics in the same row to link back and forth. You simply insert <topicref> elements in the rows of the table, as shown in Figure 7.19.

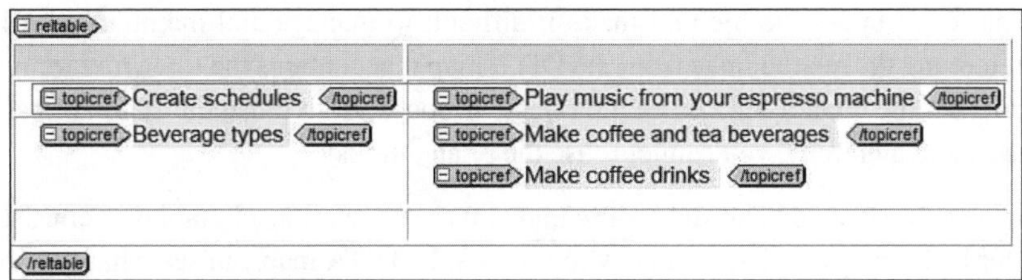

Figure 7.19 A relationship table that uses the default two-way linking.

In this example, the "Create schedules" and "Play music from your espresso machine" topics include a related link to each other in the HTML output as shown in Figure 7.20.

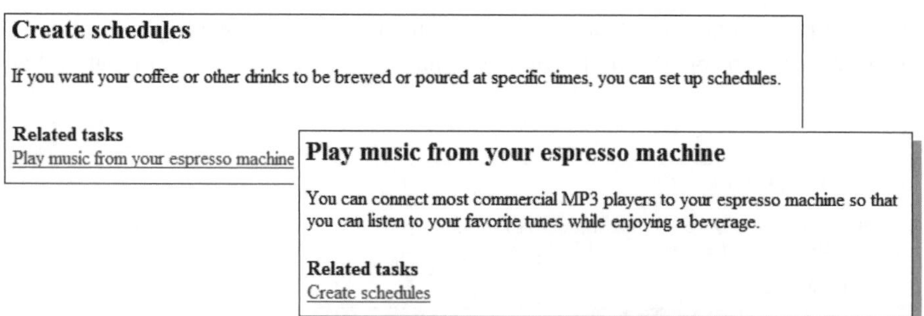

Figure 7.20 Output example from a relationship table that uses the default two-way linking.

You can also use relationship tables to link to external content. See the "External Links" section in this chapter for information about how to create related links to external websites and files.

Implementing Relationship Tables

You can include relationship tables for your information set in different ways. Don't feel tied to having one large relationship table for the information for an entire product. Consider the following options for creating relationship tables and identify a process that works best for your organization and product documentation:

- **Include one relationship table per DITA map.** The advantage is that all of your related links for topics in that DITA map are grouped in one place. The disadvantage is that you must manage links between DITA maps and decide which DITA map the links belong in. For example, consider where you should place related links between topics in a DITA map for installing the product software and a DITA map for upgrading the product software.
- **Create relationship tables in a DITA map that's separate from the topic hierarchy.** The advantage is that all of your linking code is in one file. The disadvantage is that this can result in a large file and make it difficult to manage and maintain your links. By removing the related links from the DITA map that contains the topic hierarchy, you disconnect the related links from the topics. If you want to reuse the submap that contains the topic hierarchy, you cannot reuse the related links.

Whichever format you choose, ensure that you create relationship tables in consistent locations in your DITA maps so that anyone who reviews the DITA map can see what type of linking is set up.

Don't get carried away with related links. Take advantage of the default hierarchical linking before you add related links. Hierarchical linking is nearly error-proof and low maintenance. If you decide the hierarchical linking doesn't provide enough links for your information, use related links. However, remember that too many links can be overwhelming for users.

Regardless of the type of relationship table that you use, follow these guidelines when you create links in relationship tables:

- Add links for sibling topics if you set the linking attribute to "none" in the parent topic but want to have related links for a select few of the child topics.
- Add links between topics that would not otherwise be there in a hierarchy of links.
- Add links to topics outside of the hierarchy in the current DITA map.
- Keep the number of related links in each topic to fewer than five when possible. Too many links are confusing and overwhelming.

BEST PRACTICE Avoid using the <related-links> element in a topic. If you include the use the <related-links> element to create a link in the topic, you reduce the reuse potential of the topic and make maintenance more difficult. Instead, use relationship tables.

Collection Types

A *collection type* defines a relationship among nested topics. It also specifies the way links between parent, child, and sibling topics can appear in the output.

Create a collection of linked topics by setting the value of the collection-type attribute, which defines how the nested topics are related. For example, you can specify a collection-type attribute to connect a series of task topics.

You can apply the collection-type attribute on:

- <topicref> elements in a DITA map
- <topicgroup> elements in a relationship table
- Columns in relationship tables

Apply the collection-type attribute to a parent topic reference (<topicref> element). You can create a collection type at any level in the DITA map if you have a parent topic that includes several child topics. Setting the collection-type attribute on a child topic that has no nested topics has no effect on the output, as shown in Figure 7.21.

To specify a collection type, set the collection-type attribute to one of the values shown in Table 7.2.

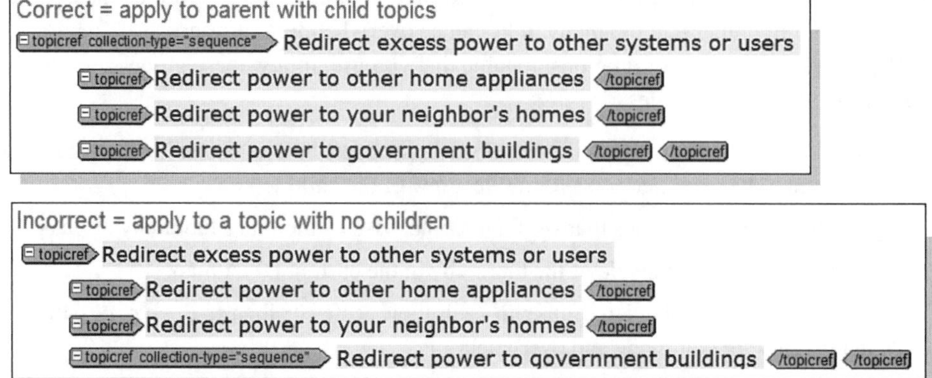

Figure 7.21 Apply the collection-type attribute only to parent topics.

Table 7.2 Collection-Type Values

Collection Type Value	Description
Sequence	Creates numbered child topics with links to the next topic and the previous topic
Choice	Creates unordered child topics
Unordered	Creates unordered child topics
Family	Creates unordered child topics and related links between the child topics

The collection-type attribute applies to only one level of nested topics. Topics that are nested at deeper levels are not affected.

Sequence Collection Type

When you set the collection type to "sequence" on a parent topic reference, you create a numbered sequence of the child topics. The parent topic will contain a numbered list of links to its child topics. In addition, the child topics will include the following links:

- A link to the parent topic (hierarchical link)
- A link to the previous sibling topic
- A link to the next sibling topic

By default, in the output of the parent topic, the numbered child topics will also show their short description under each linked title.

Set the collection type to "sequence" to create supertasks that have subtask, or child, topics. A *supertask* topic is a parent task topic that describes the overall task flow, and, through links, shows users what tasks they need to complete in what order.

For example, users of the Exprezzoh 9000N must complete several preparation tasks, which are described in separate topics, before they install the espresso machine. Therefore, you can set the value "sequence" on the collection-type attribute to create a supertask with subtasks (child topics) to guide users through preparing for the installation as shown in Figure 7.22.

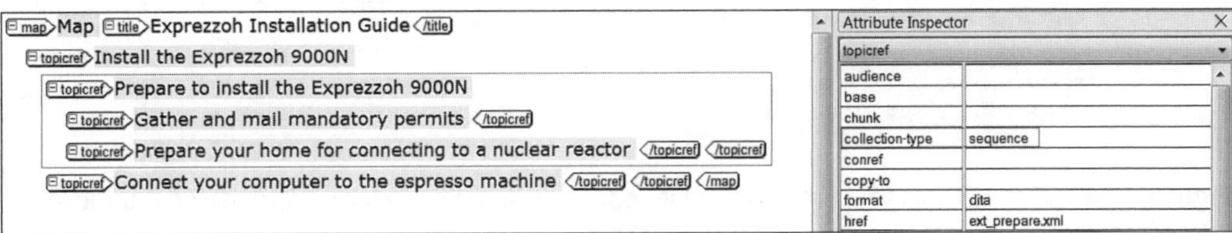

Figure 7.22 A supertask in a DITA map that uses the "sequence" collection-type.

In the HTML output for the parent topic, links to the subtask topics appear as an ordered list of steps, as shown in Figure 7.23.

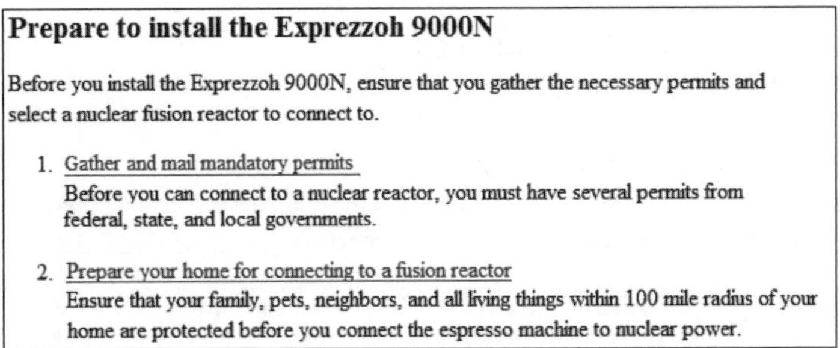

Prepare to install the Exprezzoh 9000N

Before you install the Exprezzoh 9000N, ensure that you gather the necessary permits and select a nuclear fusion reactor to connect to.

1. Gather and mail mandatory permits
   Before you can connect to a nuclear reactor, you must have several permits from federal, state, and local governments.
2. Prepare your home for connecting to a fusion reactor
   Ensure that your family, pets, neighbors, and all living things within 100 mile radius of your home are protected before you connect the espresso machine to nuclear power.

Figure 7.23 A parent topic with numbered child topics created by setting the collection-type attribute to "sequence."

The output of the first child topic in the sequence, "Gather and mail mandatory permits," includes links to both the parent topic, the previous topic, and the next sibling topic "Identify a nuclear reactor to connect to" as shown in 7.24.

By using the "sequence" collection type, you don't need to use <xref> elements to create inline links from the parent topic to the child topics or between the sibling topics.

Once again, be careful about duplicating links. You can easily forget about an inline link that you added a year ago and then later, inadvertently create a link to the same topic by setting the collection-type attribute on a <topicref> element.

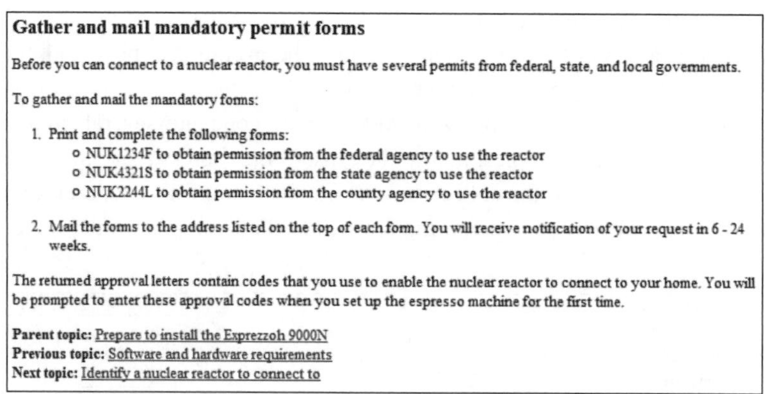

Gather and mail mandatory permit forms

Before you can connect to a nuclear reactor, you must have several permits from federal, state, and local governments.

To gather and mail the mandatory forms:

1. Print and complete the following forms:
   - NUK1234F to obtain permission from the federal agency to use the reactor
   - NUK4321S to obtain permission from the state agency to use the reactor
   - NUK2244L to obtain permission from the county agency to use the reactor
2. Mail the forms to the address listed on the top of each form. You will receive notification of your request in 6 - 24 weeks.

The returned approval letters contain codes that you use to enable the nuclear reactor to connect to your home. You will be prompted to enter these approval codes when you set up the espresso machine for the first time.

Parent topic: Prepare to install the Exprezzoh 9000N
Previous topic: Software and hardware requirements
Next topic: Identify a nuclear reactor to connect to

Figure 7.24 A child topic that links to both its parent, the previous topic, and the next topic in the sequence.

Avoid manually creating link lists to "fake" a collection of topics if you can simply nest those topics in a DITA map. In the following example in Figure 7.25, although child topics are nested in the DITA map to create a supertask hierarchy, the parent topic also has inline links to the child topics.

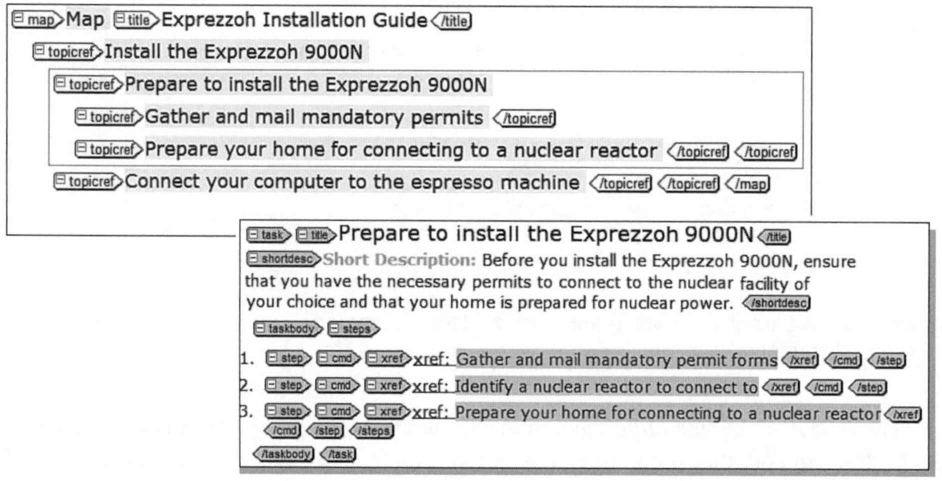

Figure 7.25 A "fake" supertask created in a parent topic by using cross references to child topics.

In HTML output, you can see the duplicate links that are created from the sequence collection type and inline links, as shown in Figure 7.26.

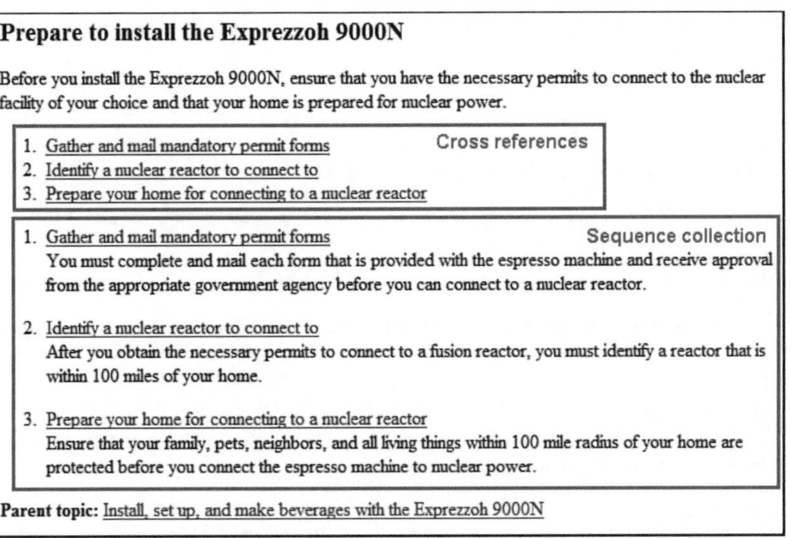

Prepare to install the Exprezzoh 9000N

Before you install the Exprezzoh 9000N, ensure that you have the necessary permits to connect to the nuclear facility of your choice and that your home is prepared for nuclear power.

Cross references

1. Gather and mail mandatory permit forms
2. Identify a nuclear reactor to connect to
3. Prepare your home for connecting to a nuclear reactor

Sequence collection

1. Gather and mail mandatory permit forms
   You must complete and mail each form that is provided with the espresso machine and receive approval from the appropriate government agency before you can connect to a nuclear reactor.
2. Identify a nuclear reactor to connect to
   After you obtain the necessary permits to connect to a fusion reactor, you must identify a reactor that is within 100 miles of your home.
3. Prepare your home for connecting to a nuclear reactor
   Ensure that your family, pets, neighbors, and all living things within 100 mile radius of your home are protected before you connect the espresso machine to nuclear power.

Parent topic: Install, set up, and make beverages with the Exprezzoh 9000N

Figure 7.26 Duplicate links in HTML output created by inserting <xref> elements in the topic "Prepare to install the Exprezzoh 9000N" and by using a sequence collection type.

Choice Collection Type

Use the "choice" collection type when you want to create a supertask that presents subtask topics as choices. For example, if a supertask topic provides an overview of how to make a beverage with the espresso machine, the subtask topic choices might be:

- Make coffee beverages
- Make hot tea
- Make soda or juice beverages
- Make alcoholic beverages

To set up this collection, you specify "choice" for the collection-type attribute on the parent topic <topicref> element, as shown in Figure 7.27.

topicref Make coffee, tea, soda, and alcoholic beverages
topicref Make coffee beverages
topicref Make hot tea /topicref
topicref Make soda and juice beverages /topicref
topicref Make alcoholic beverages /topicref /topicref

audience	
base	
chunk	
collection-type	choice
conref	
copy-to	
dir	

Figure 7.27 Nested child task topics defined as a "choice" collection type.

By default, the HTML output shows a simple list of links to the child topics with no bullets as shown in Figure 7.28.

Make coffee, tea, soda, and alcoholic beverages

The Exprezzoh 9000N is more than just an espresso drink maker: It can pour or mix tea, soda, and even alcoholic beverages.

Ensure that you have the ingredients for the beverages that you want. Ingredients for beverage types are sold separately.

You can create many types of espresso drinks, such as espresso shots, cappuccino, machiato, latte, and mocha. You can also create customized drinks by running commands from your computer and importing those instructions to the espresso machine.

> Make coffee beverages
> With the Exprezzoh 9000N, you can make delicious coffee and tea beverages by using the default grind and temperature settings or by customizing the brew to your tastes.
>
> Make hot tea
> You can make tea simply by dispensing hot water from the espresso machine and adding a packaged tea, or you have the espresso machine steep loose-leaf tea for specific time.
>
> Make soda and juice beverages
> You can dispense sodas or juices from the espresso machine. You can also have the espresso machine make you mixed drinks, such as lemon-lime soda mixed with orange juice.
>
> Make alcoholic beverages
> The espresso machine can dispense and mix various alcoholic beverages. You must provide the beverages and for mixed drinks, use the dashboard to devise your own perfect drinks.

Figure 7.28 Nested child task topics in HTML output defined as a "choice" collection type.

TIP You can modify your style sheet to display the list of child topics as a bulleted list instead of a simple list with no bullets.

If you select "choice" as a collection type, the parent and child topics have the same default behavior as hierarchical linking. That is, you see child topics under a parent topic, but you won't get linking between child topics, unlike the "sequence" collection type.

Unordered Collection Type

Select "unordered" as a collection type if the procedures in the child topics can be completed in any order. Just like the "choice" collection type, you'll see child topics under a parent topic, but you won't get linking between child topics (see Figure 7.29).

In HTML, the nested topics appear as a simple list of links to the child topics as shown in Figure 7.30.

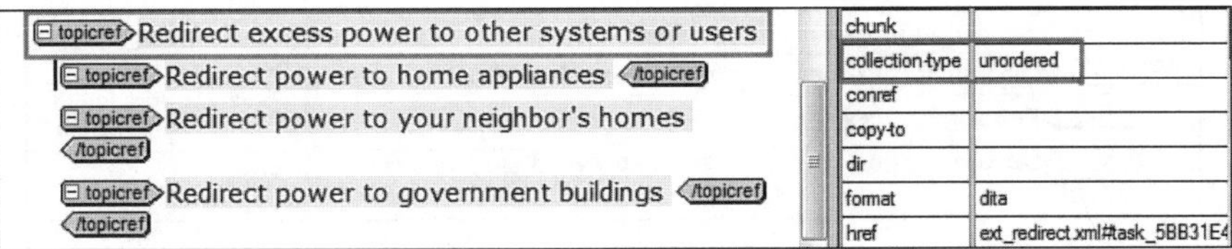

Figure 7.29 Nested child task topics in HTML output defined as an "unordered" collection type.

Redirect excess power to other systems or users

Even though the espresso machine requires the incredible power generated from nuclear fusion, sometimes all that energy is not needed. You can redirect excess energy to other appliances in your home, your neighbors, or city and county buildings.

> **Redirect power to home appliances**
> You can use the excess energy from your espresso machine to power appliances such the washer and dryer, guitar amplifiers, and televisions.
>
> **Redirect power to your neighbor's homes**
> If you have excess energy from your nuclear power source, you can offer that energy to other homes within a one square mile of your home.
>
> **Redirect power to government buildings**
> You can redirect excess power from your nuclear energy to government buildings. The administrators of those buildings must participate in the Nuclear Energy Sharing program.

Parent topic: Monitor the espresso machine system

Figure 7.30 Nested child task topics in HTML output defined as an "unordered" collection type.

TIP You can modify your style sheet to display the list of child topics as a bulleted list instead of a simple list. Consider whether you need to distinguish the display of child topics in a choice collection type from the display of child topics in an unordered collection type.

Family Collection Type

The "family" collection type is similar to the "unordered" and "choice" collection types, but the "family" collection type also includes related links to sibling topics, unlike the "unordered" and "choice" collection types.

With the "family" collection type, you get the following linking in the output:

- Child topics link to their parent topic (hierarchical links).
- The body of the parent topic contains links to each child topic as an unordered link list.
- Each child topic links to its sibling topics as related links.

In Figure 7.31, the collection-type attribute is set to "family," which creates two child links.

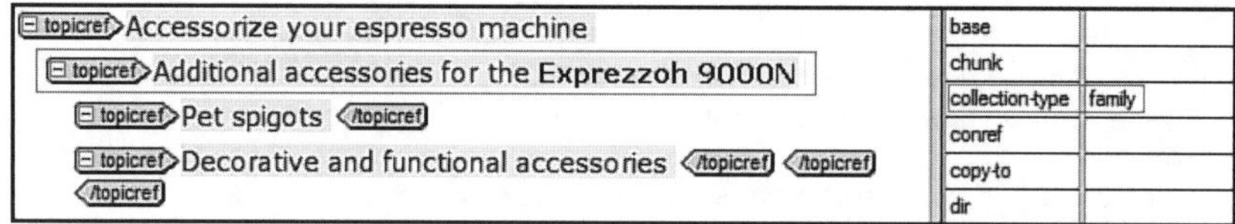

Figure 7.31 Nested child task topics in HTML output defined as a "family" collection type.

In HTML output, the linked child topics appear at the end of the parent topic content, as shown in Figure 7.32.

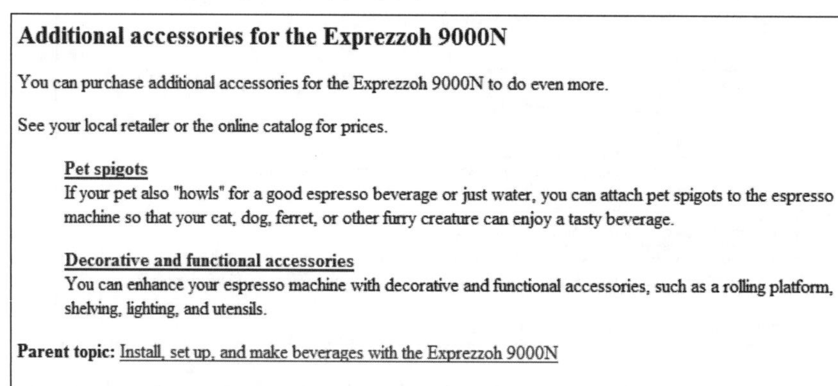

Additional accessories for the Exprezzoh 9000N

You can purchase additional accessories for the Exprezzoh 9000N to do even more.

See your local retailer or the online catalog for prices.

Pet spigots
If your pet also "howls" for a good espresso beverage or just water, you can attach pet spigots to the espresso machine so that your cat, dog, ferret, or other furry creature can enjoy a tasty beverage.

Decorative and functional accessories
You can enhance your espresso machine with decorative and functional accessories, such as a rolling platform, shelving, lighting, and utensils.

Parent topic: Install, set up, and make beverages with the Exprezzoh 9000N

Figure 7.32 A parent topic with two child topic links created by setting the collection-type attribute to "family."

In the HTML output as shown in Figure 7.33, the child topics also include related links to each other, which in this example, means that the sibling topic "Decorative and Functional Accessories" links to its sibling topic "Pet spigots."

Determining Which Collection Type to Use

To determine which collection type value to use, answer the questions in Table 7.3.

Table 7.3 How to Determine Which Collection Type to Use

Question	Collection-Type Value
Must the nested task topics be done in a particular order?	Sequence
Do the topics represent choices?	Choice or Family
Can the task topics be done in any order?	Unordered or Family
Are the topics closely related and need links between sibling topics?	Family

Decorative and functional accessories

You can enhance your espresso machine with decorative and functional accessories, such as a rolling platform, shelving, lighting, and utensils.

Accessory	Part number
Attachable lighting	EXP9000AL
Foam spatula	EXP9000FS
Large umbrella	EXP9000LU
Set of 4 coffee mugs (8 oz.)	EXP9000CM
Set of 4 espresso cups (4 oz.)	EXP9000EC
Stainless steel milk foamer cup	EXP9000MF

See your local retailer or online catalog for ordering information.

Parent topic: Additional accessories for the Exprezzoh 9000N

Related reference
Pet spigots

Figure 7.33 A child topic that links to its sibling topic under the heading "Related reference."

Collection Types in Relationship Tables

You can use collection types in relationship tables by setting the value of the collection-type attribute on:

- The <topicgroup> element for a group of topic references in the same cell
- The <relcolspec> element for all topic references in that column

Setting Family Collection Types

You can set the collection-type attribute in your relationship table to "family" to create related links between topics.

The following example shows the collection-type attribute set to "family" on a <topicgroup> element that contains topic references in a two-way relationship table. This creates two-way links between the topics in the <topicgroup> element, as shown in Figure 7.34.

You can also use a single-column relationship table and set the collection-type attribute for the column to "family."

The relationship table in Figure 7.35 creates two-way links between the topics in the same cell.

Setting Sequence Collection Types

Often times you want to get the output of a sequenced collection type, but the topics aren't organized appropriately in the DITA map, and therefore, you can't use the "sequence" collection type.

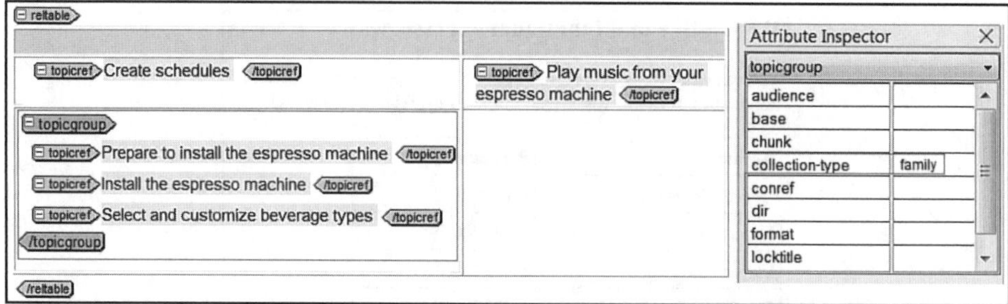

Figure 7.34 A collection-type attribute with the value "family" in a relationship table.

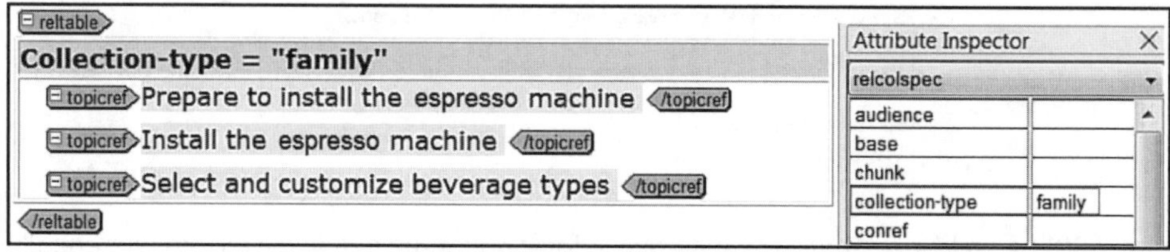

Figure 7.35 A one-column relationship table that uses the collection type of "family."

To get around this problem, you can create a sequenced collection type in a relationship table to define sequential links that are independent of the topic hierarchy in the DITA map.

For example, if you want to show the main flow of an installation procedure that requires several task topics, you can create a supertask parent topic that has no child topics nested in it in the DITA map hierarchy. Instead, add the supertask parent topic to a cell in a relationship table and nest the child topics in the order that you want them to appear in the output. Set the collection-type attribute on the parent <topicref> element to "sequence" as shown in Figure 7.36. This is a great way to show high-level workflows and tasks without having to create a list of inline links to the target topics.

Setting up related links this way provides the HTML output shown in Figure 7.37, which looks the same as nesting child topics in the DITA map.

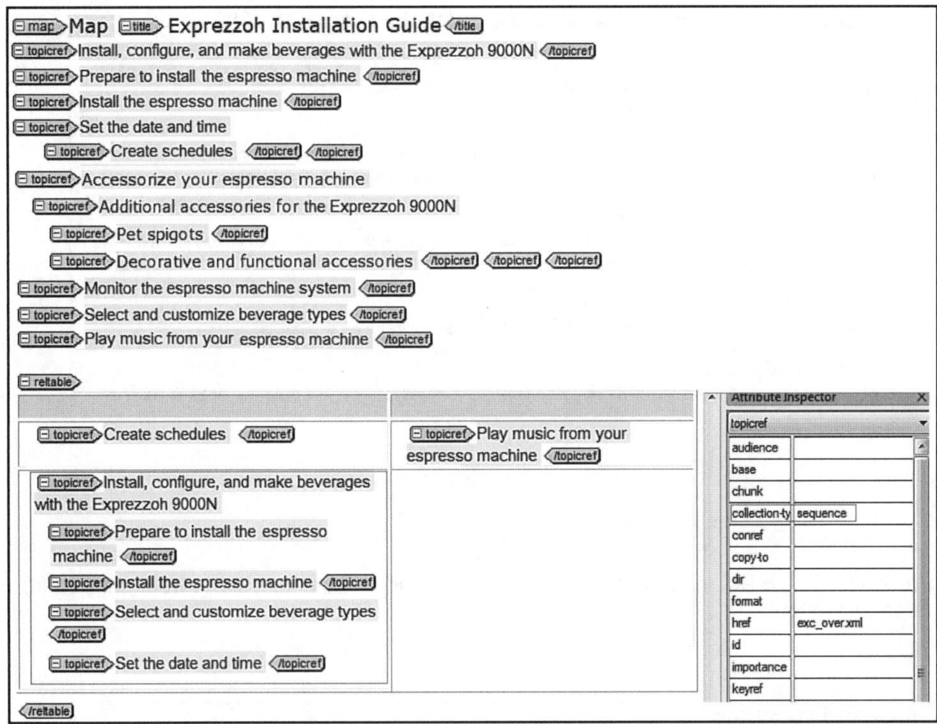

Figure 7.36 A relationship table with the value of "sequence" set in a relationship table cell.

Links Created with the Importance Attribute

Instead of manually adding an inline link in the prerequisite section of a task topic with an <xref> element, you can have a link automatically inserted in the prerequisite section of a task topic through the magic of DITA.

OK, maybe not magic, but adding such links is simple. To create a prerequisite link, set the collection-type attribute on the parent <topicref> element to "sequence." Then, set the importance attribute on the prerequisite topic's <topicref> element to "required" as shown in Figure 7.38.

When you create HTML output, the topic with the attribute set to "required" displays as a prerequisite link at the top of all other child topics in that sequence of topics, as shown in Figure 7.39.

WATCH OUT Your output won't be consistent if you use this method to automatically insert prerequisite links in some topics but add descriptions of prerequisite information without links in other topics. Be sure to establish guidelines for linking to topics that contain prerequisite information.

Install, configure, and make beverages with the Exprezzoh 9000N

The Exprezzoh 9000N does more than just help you make coffee beverages. This scalable, cross-dimensional product can display calorie, caffeine, and other nutrition facts, feed your pets, power most other appliances in your home, play your favorite music, and do many other tasks.

The Exprezzoh 9000N is a powerful household tool because it uses nuclear energy as its power source. Just hook it to a nuclear fusion reactor and you are ready to make great coffee and other beverages.

1. Prepare to install the Exprezzoh 9000N
   Before you install the Exprezzoh 9000N, ensure that you gather the necessary permits and select a nuclear fusion reactor to connect to.
2. Install the espresso machine
   You can easily connect the nuclear fusion reactor to the coffee maker to install the espresso machine.
3. Select and customize beverage types
   To have the Exprezzoh 9000N server beverages, you must first select and optionally customize beverage types.
4. Set the date and time
   After you install the espresso machine, you must set the date and time.
5. Monitor the espresso machine system

Figure 7.37 The HTML output of a supertask and its sequential topics that were created from the relationship table.

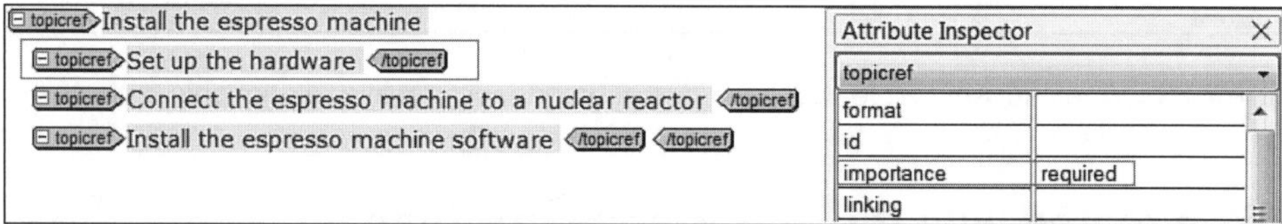

Figure 7.38 Nested topics with one topic's importance attribute set to "required."

Linking Scope

The scope attribute of an <xref> or <topicref> element indicates the location of a source topic relative to the location of a target topic. You set the scope attribute differently depending on where the target content is in relation to the source topic. For example, the target link might go to a sibling topic in the same DITA map, or it might go to an external website.

You must specify the scope attribute for links that you create by using the <xref> or <topicref> elements.

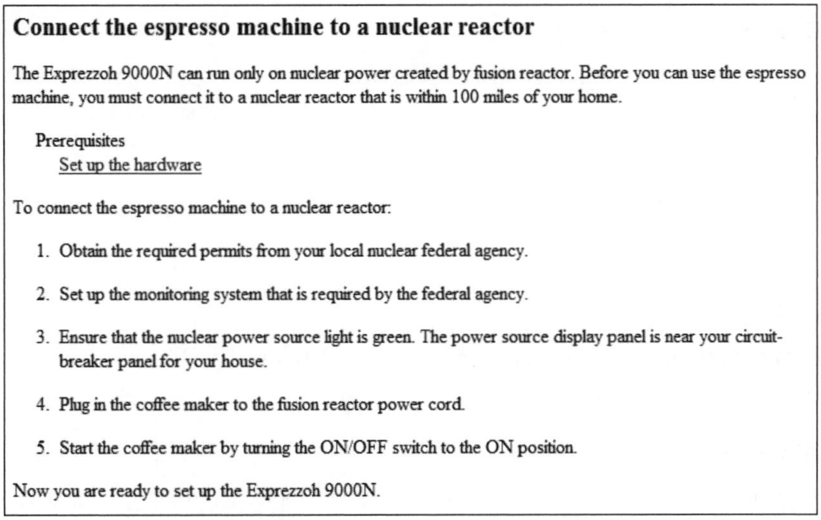

Connect the espresso machine to a nuclear reactor

The Exprezzoh 9000N can run only on nuclear power created by fusion reactor. Before you can use the espresso machine, you must connect it to a nuclear reactor that is within 100 miles of your home.

Prerequisites
Set up the hardware

To connect the espresso machine to a nuclear reactor:

1. Obtain the required permits from your local nuclear federal agency.
2. Set up the monitoring system that is required by the federal agency.
3. Ensure that the nuclear power source light is green. The power source display panel is near your circuit-breaker panel for your house.
4. Plug in the coffee maker to the fusion reactor power cord.
5. Start the coffee maker by turning the ON/OFF switch to the ON position.

Now you are ready to set up the Exprezzoh 9000N.

Figure 7.39 A prerequisite link that is automatically added to sibling topics.

You typically link from a source topic to target content in one or more of these situations:

- Linking to a topic that is in the set of topics that you own and maintain together. For example, you link between two topics in your installation guide.
- Linking to a topic that other writers own but isn't delivered with your set of topics. For example, you link from a topic in your installation guide to a topic in the upgrade guide.
- Linking to an external website. For example, you link to your company support page.

Depending on which situation you're in, use the following values to set the scope attribute:

- **Local:** Link to topics that are currently available to the set of topics when you build output of the information set.
- **External:** Link to content that is outside the set of topics that you're building output.
- **Peer:** Link to content that might not be available when you build output of your set of topics but will be available when the information is delivered to users.

Figure 7.40 shows the scope attribute set to "local" on an <xref> element that links the current task topic to a task topic that is available in the same information set.

Depending on your XML editor, the scope value might already be set for you based on the file extension in the href attribute of the <xref> element. For example, a reference to a DITA file might have the default value of "local" whereas a reference to an HTML file might have the default value of "external."

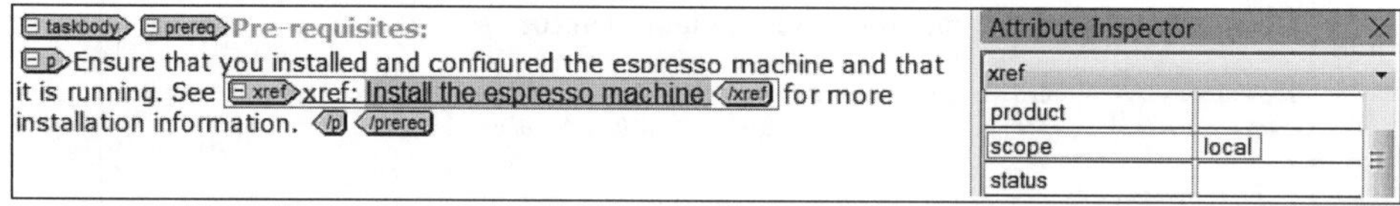

Figure 7.40 Setting the scope attribute of an <xref> element.

Local Links

The majority of your links will probably be local links. Use the "local" scope for links to topics that are available in your set of topics.

External Links

When you create an external link in a topic or relationship table, set the scope attribute to "external," and include the complete URL to the file or site that you're linking to.

For example, if you want to include an inline link in a topic to the website of the fictional product Exprezzoh at http://www.exprezzoh.com, add an <xref> element and the following attribute values:

```
<xref href="http://www.ezprezzoh.com" scope="external"
  format="html">www.ezprezzoh.com</xref>
```

The format value specifies the type of target content that you are linking to. For example, if you link to websites, specify "html" for the format attribute.

TIP Create guidelines that define when it's appropriate to expose the URL or to hide the URL and add link text for external hyperlinks. If users print your information, they might need to see the URL. If the website that you are referring to is easy to find on the Internet or changes often, you should write text for the hyperlink instead and rely on your users' ability to search for the website by using a search engine. For example, you can use "Exprezzoh Espresso Machines" for the link text rather than "/www.exprezzoh.com."

We recommend that you always set the link text and add short descriptions for external links included in relationship tables. Remember that the short description is displayed when a user hovers over a link in HTML output.

To create an external link in a relationship table, you add the link text and short descriptions by inserting the <topicmeta>, <linktext>, and <shortdesc> elements in the <topicref> element as shown in Figure 7.41.

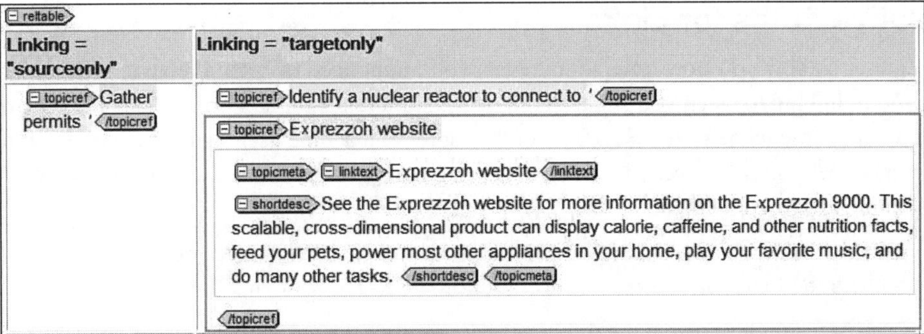

Figure 7.41 Link text and a short description added to a relationship table for an external link.

External related links are labeled as "Related information" in the output. When users hover over the external related link in the HTML output, the short description from the relationship table is displayed, as shown in Figure 7.42.

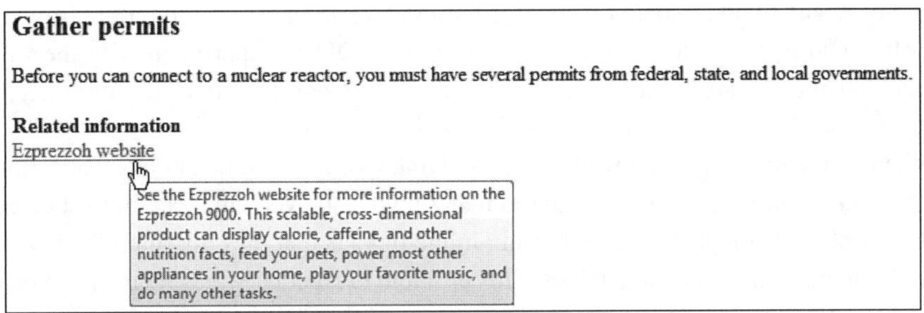

Figure 7.42 Link text and a short description added to a relationship table for an external link as shown in HTML output.

Peer Links

Most of the time, you'll set the link scope to "local" or "external." However, occasionally, you might need to link to topics that aren't yet available or that exist in a separate component of your information set. In these rare cases, set the link scope to "peer" to indicate that the topic is not available at build time but will be available when all of the output is combined and delivered to users. If you do not set the peer link's scope appropriately, you will get broken links and transform errors.

Relative Paths for Links

Depending on how you store your source files, you might need to enter a relative path to the file that you're linking to when you create an inline link or a related link. For example, to link from a topic with the file name `exp_use.dita` in the espresso machine user's guide to a topic with the file name `exp_install.dita` in the installation guide, you can create the following <xref> element in the exp_use.dita topic:

```
<xref href="../InstallGuide/exp_install.dita"  scope="peer">
  Installing the Exprezzhoh 9000N</xref>
```

The "`../`" indicates that the link is relative. In your source control system, the DITA topic that's being linked to is one directory above the folder named `InstallGuide`.

Link Testing

How frustrated are you when you click to find a broken link? Don't risk your company's reputation and user satisfaction by allowing broken links to live in your information. Testing your links should be part of your production process. Build a link checking procedure into your process to identify broken links, incorrect links, unnecessary links, or duplicate links.

The best way to understand how your linking is working is to build output of your information often and test the links. As you learn to link in DITA, experiment with the values for the various attributes and build output to see how those settings affect the usability of your content.

When you combine automatically generated links from hierarchical links, collection types, and relationship tables, you might create too many links and even duplicate links. For example, one of the most common problems with using inline links is that they are often duplicated when you also create links in relationship tables. This problem is especially prevalent when you convert content to DITA and the original content contains many inline links.

For example, Figure 7.43 contains an inline link to the topic called "Make Hot Tea," but it also contains the same link under "Related tasks." The inline link is created by using an <xref> element inserted in the body of the topic. The "Related tasks" link is created by using a relationship table.

When possible, use relationship tables to create links instead of inserting inline links so that you can avoid duplicate links. Maintaining links in relationship tables is often easier than tracking inline links in topics.

To Wrap Up

Your topic-based content won't be complete and useful until topics are properly linked to other related topics. With the right links, your users can easily follow a task flow and find relevant topics to help them complete their goals.

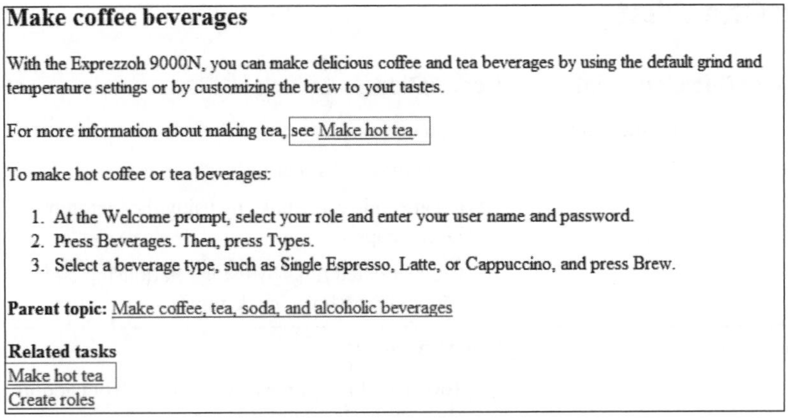

Make coffee beverages

With the Exprezzoh 9000N, you can make delicious coffee and tea beverages by using the default grind and temperature settings or by customizing the brew to your tastes.

For more information about making tea, see Make hot tea.

To make hot coffee or tea beverages:

1. At the Welcome prompt, select your role and enter your user name and password.
2. Press Beverages. Then, press Types.
3. Select a beverage type, such as Single Espresso, Latte, or Cappuccino, and press Brew.

Parent topic: Make coffee, tea, soda, and alcoholic beverages

Related tasks
Make hot tea
Create roles

Figure 7.43 Duplicate links to the topic called "Make hot tea."

You can use the following types of links to connect your topics into a coherent web of information:

- Hierarchical links
- Inline links
- Related links
- Collection-type links

Proper linking also helps both advanced and novice users:

- Novice users can follow links to get more information.
- Advanced users can skip links they don't need.

Be careful not to get carried away with linking: Too many links can be just as bad as not having enough. You must find the right balance between not stranding users in dead-end topics and providing enough relevant related links so that users don't get lost. Remember that duplicate links are confusing, too many links are overwhelming, links to the wrong topics are unhelpful, and having no links when you need them is frustrating.

Finally, decide on a linking strategy with your team and evaluate the effectiveness of the linking in your editing and quality assurance processes. Effective linking will improve the quality of your information.

Linking Checklist

Guideline or Decision Point	Description
Use the right linking for the right situation.	Remember to: • Use inline links sparingly. • Create child topic links by using the hierarchical linking in DITA maps. • Create related links by using relationship tables, not by adding inline links inside topics.
Be careful with link sprawl.	Remember that: • Too many links are overwhelming. Avoid adding more than about five related links to a topic. • The more links you have in your information, the less important each link becomes.
Build your information by starting with the default linking in DITA.	Decide whether default hierarchical linking provides the links that you need. If you need more links, use relationship tables and collection types. Lastly, insert inline links if these links aren't adequate.
Use collection types correctly.	Organize your topics by setting the following collection types: • Choice • Family • Sequence • Unordered
Provide link text and short descriptions for most external links.	If your information will be printed, consider showing the full URL rather than using text for links. However, in most cases for external links, follow these guidelines: • Don't just provide an ugly URL for external links. Instead, use text that explains where the link is going. • Add a short description that briefly describes the content that the link goes to. • Be consistent with the link text and short descriptions for external links. For example, don't include short descriptions for some external links and not for others.

Guideline or Decision Point	Description
Consider using the importance attribute to create prerequisite links.	Instead of manually adding an inline link in the prerequisite section of a task topic, insert a link automatically in the prerequisite section of a task topic by setting "required" on the importance attribute of the <topicref> element. Be consistent in the way that you link or describe prerequisite information.
Understand the values of the scope attribute.	Use the following values of the scope attribute to indicate the location of source topics relative to the location of the target topic: • Local • External • Peer
Implement a link checker to automate quality assurance testing of your linking strategy.	Testing your links should be part of your production process. Build a link checking procedure into your process to identify broken links, incorrect links, unnecessary links, or duplicate links.

CHAPTER 8

Metadata

We'll be honest: Dealing with some areas of metadata is for the more advanced DITA user. If your writing team is just learning about DITA elements, don't scare them by using fancy words such as metadata at team meetings. Otherwise, the guy who brings the donuts might not come anymore.

Fortunately, it's not all rocket science. For example, metadata also includes index entries, which are easy to implement.

Metadata is information about the content in your DITA topics and maps. Properly implemented, metadata can help writers create content targeted to specific audiences, products, versions, and so on, which in turn helps your users to find the right information.

DITA includes many elements and attributes in DITA maps and topics that are specifically for providing metadata. You can use these metadata elements to:

- Automatically include information to meet legal or company requirements.
- Improve the retrievability of your content.
- Customize your information for different audiences, versions, models, and so on.

Why Is Metadata Important?

Metadata is one of the best ways to improve the retrievability of your information for both users of your products and for your writers. It won't matter how accurate, complete, or grammatically correct your topics are if users can't find the information that they need.

A good metadata strategy can help you create information that is:

- **Easy to find:** Metadata, such as keywords and index entries, helps search engines and users find specific content.
- **Easy to manage:** Content management tools can use metadata to help you find and organize topics and DITA maps that are properly classified with metadata.
- **Targeted to a specific audience:** You can write one set of topics but include or exclude information based on version, model, operating system, audience, and so on to ensure that specific users see only the information that they need.

Metadata can also help you to implement cutting-edge content delivery features such as dynamic publishing and faceted browsing. For example, you might set up a publishing mechanism so that users can select the types of content that they need based on categories such as operating system or server types for software products, or model number or size for hardware products.

Faceted browsing helps to narrow the number of topics that are relevant to a specific user by only displaying the content that is applicable to what they are searching for. For example, some users might want to see only installation topics that apply to the Windows operating system, as shown in Figure 8.1. Those topics might include more background information designed for the novice user.

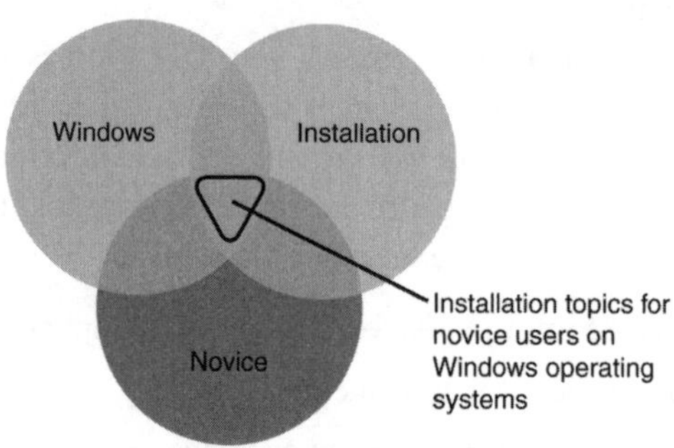

Figure 8.1 Metadata that can help users to filter content.

Figure 8.2 shows how users of the Exprezzoh 9000N product might select or filter topics that are appropriate for them in the help system.

Types of Metadata

DITA has several types of metadata. Depending on your content, you might use some or all these types of metadata:

- Index entries
- Conditional processing attributes

- Topic metadata
- DITA map metadata

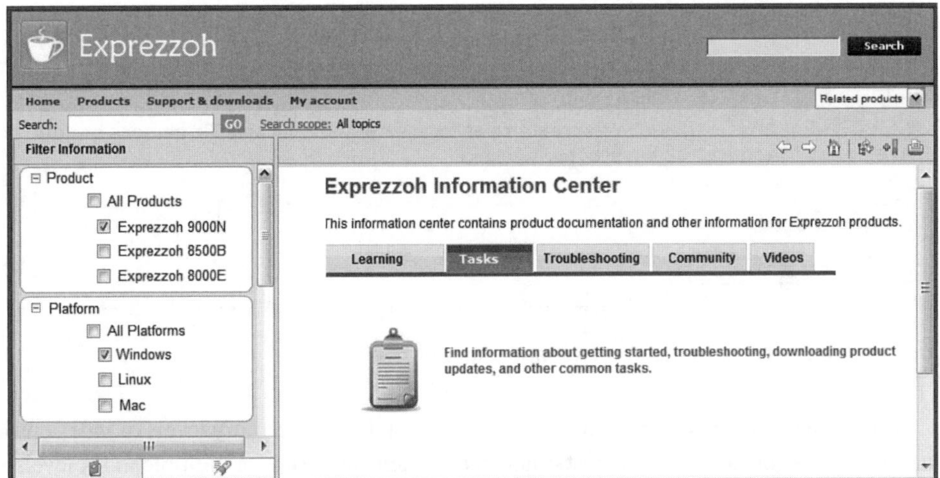

Figure 8.2 Metadata that is the basis of faceted browsing in a help system.

BEST PRACTICE Apply metadata in your topics and DITA maps consistently and as you create content rather than wait until you finish writing topics. Otherwise, you'll get inconsistent and unexpected results when you build your output.

Index Entries

Let's start with the easier of the metadata types: index entries.

Index entries are a critical form of metadata because they help to improve the retrievability of content. Index entries are used to create an index for the information set in PDF and HTML output, but when was the last time that anyone used an index? Indexes are obsolete because everyone uses a search tool to find information these days, right? Wrong!

Index entries can also be used as search terms for external web crawlers. Even if no one ever sees your index, the index entries can help users find your information.

In DITA, you can place index entries almost anywhere in a topic, but you should insert most of them in the <keywords> element of the <prolog> element. This ensures that the index entries in each topic are in a single location and that the terms can best be optimized for search engines.

If you insert index entries in the <prolog> element, when users click an entry in the index, they are taken to the beginning of the topic. If you stick to good topic-writing guidelines and keep your topics short and focused, you shouldn't need to add index entries outside of the <prolog> element.

Use the <indexterm> element to add index entries. Figure 8.3 shows an index term included in the <prolog> element.

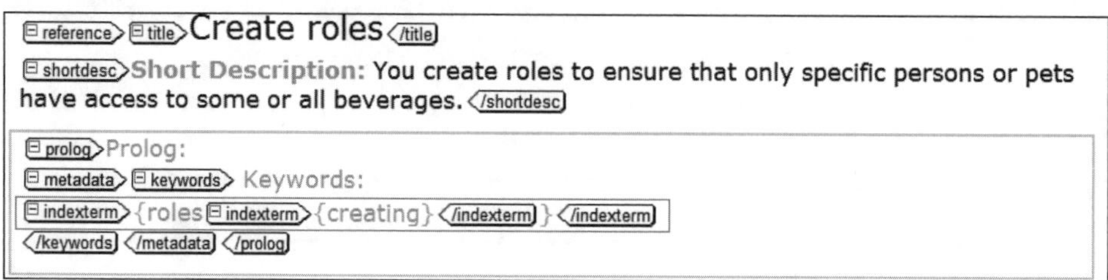

Figure 8.3 Index terms in the <prolog> and <keyword> elements.

You can nest <indexterm> elements to create secondary and tertiary levels of index entries. Figure 8.4 shows you how you can nest <indexterm> elements to create multiple index levels.

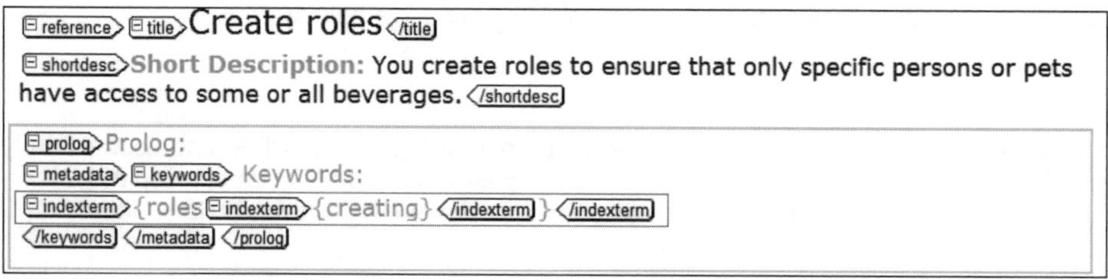

Figure 8.4 An <indexterm> element nested in another <indexterm> element to create primary and secondary index entries.

See and See Also Entries

Use See entries (<index-see> element) to direct users to a preferred term for your product. For example, if your product uses the term *fusion* instead of *thermonuclear fusion*, use the <index-see> element to direct users from one term to the other, as shown in Figure 8.5.

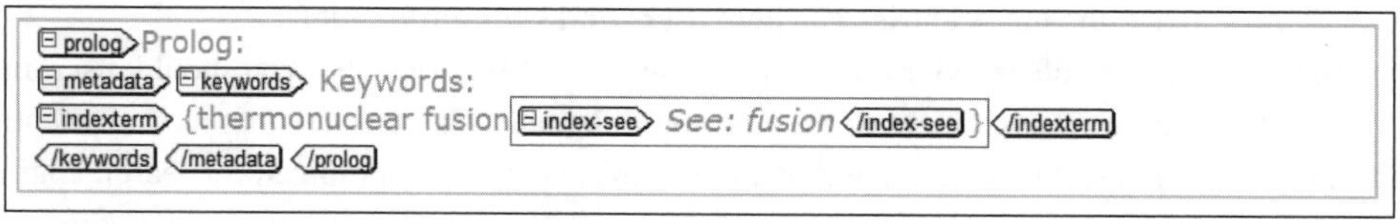

prolog Prolog:
metadata keywords Keywords:
indexterm {thermonuclear fusion index-see See: fusion /index-see } /indexterm
/keywords /metadata /prolog

Figure 8.5 A See index entry that redirects users to a preferred term.

Use See Also entries (<index-see-also> element) to direct users to related but not necessarily synonymous terms. For example, the nuclear-powered espresso machine relies on fusion energy, but users might want to compare *fusion* energy to *fission* energy. Therefore, you can use the <index-see-also> element to add a reference to *fission*, as shown in Figure 8.6.

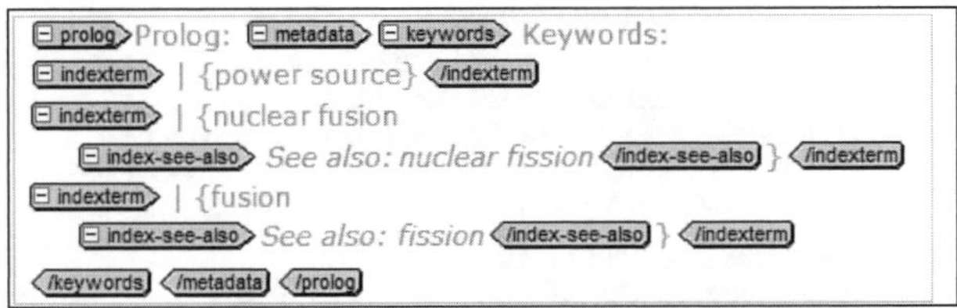

Figure 8.6 A See Also index entry that redirects users to related information.

WATCH OUT Don't overuse the See and See Also references in indexes. They are rarely needed. If you have many index entries that use a See or See Also reference, your users could become frustrated hopping around your index instead of going directly to the information that they need. Also, by default, See and See Also entries don't include page numbers in PDF output.

Index Entries for Long Topics

For long topics that are more than three pages, you might add index entries closer to the information that the index entry corresponds to, such as near tables, table rows, or definition list items rather than putting all the index entries in the <prolog> element. Placing the index entries closer to content that you want users to find ensures that the user finds the information more quickly from the index.

The following command reference topic is three pages in PDF output because it includes a long table (see Figure 8.7). Users might not easily find specific commands in the table if the index entries all point to the top of the topic. Including an index entry next to each command in the table makes it easier for users to find a specific command from the index.

The <index-sort-as> Element

If you're control freaks like we are or index lovers like our editors, you might want to control how some index terms are sorted. Use the <index-sort-as> element to specify a different sorting phrase for an index entry. For example, you might want to ignore the special character that precedes a term such as #LOADTERMINAL so that the index entry appears under L rather than under special characters at the beginning of the index.

Figure 8.8 shows the <index-sort-as> element inside a standard <indexterm> entry.

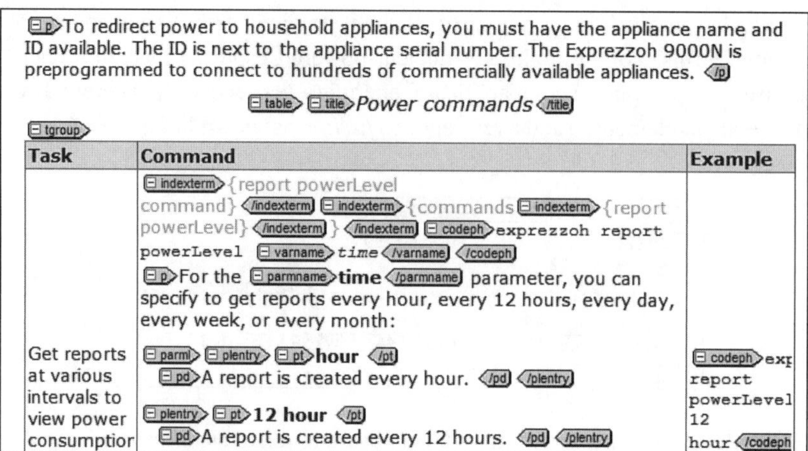

Figure 8.7 Index terms in a long table.

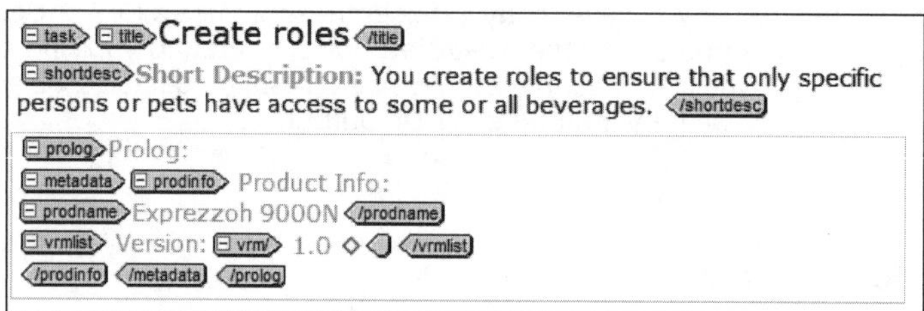

Figure 8.8 The <index-sort-as> element that sorts the term *loadterminal.*

The <index-sort-as> element can also be used by localization teams who translate content into other languages and who might need to sort terms differently than they would in English. For example, Japanese index terms are sorted phonetically as opposed to alphabetically. Work with your localization team to understand the needs of your global customers to determine whether you should use the <index-sort-as> element for localization.

Indexing Guidelines

Just as you create writing guidelines to ensure the quality of your text, create indexing guidelines to ensure that writers create consistent index entries.

Good indexing skills can be difficult to master. If you have an indexing guru on your team, buy him or her a latte every once in a while.

Tips for Writing Index Entries

Although this book doesn't provide detailed indexing advice, follow these general indexing guidelines:

- Ensure that every topic that has substantial content contains an index entry.
- Limit the nesting levels of your index terms to secondary or tertiary.
- Do not capitalize indexed terms unless they are proper nouns.
- Think nouns first, verbs second. For example, make the primary entry "databases" and the secondary entry "installing."
- Use plural nouns unless only singular makes sense. And be consistent. Don't create the entry "role" and then in another topic, create the entry "roles." Use only the plural.
- Avoid gerunds and imperative verbs, such as "installing" and "install" as primary entries unless no other nongerund noun is appropriate. For example, you can use "troubleshooting" as a primary entry.
- Cross-post (reverse) the primary and secondary entries when possible. For example, create "databases" as a primary entry and "installing" as a secondary entry. Then, create "installation" as a primary entry and "databases" as a secondary entry.
- Avoid inserting too many See and See Also entries.
- Use the <index-sort-as> element to sort unusual terms to help users find those terms in the index.
- Index the subjects in the topic, not just the topic title.

Conditional Processing Attributes

Conditional processing attributes are a key benefit of DITA—so much so that we've created another chapter dedicated to it. Chapter 9, "Conditional Processing," explains how you can apply conditional processing attributes at the topic and element level to include or exclude content or flag content with a graphical icon to make scanning your information easier.

You can also take advantage of processing attribute information in other ways. The values that you apply to conditional processing attributes, such as audience, product, platform, and rev (revision), are metadata. Remember that metadata is information about your content and that these attributes can contain lots of valuable information about your content.

Metadata that you enter for conditional processing can be used to:

- Improve the ability of writers to locate topic and DITA maps in a file storage system
- Support customized content processes, such as dynamic publishing

The members of the DITA community don't completely agree that metadata in conditional attributes should be treated differently than metadata in <metadata> and <topicmeta> elements. For example, some say that you shouldn't use the conditional processing attributes for anything but filtering and flagging content. However, we think that all metadata can be used to define your information more precisely and target that information to the right users.

Importance, Status, and Translate Metadata Attributes

In addition to conditional processing attributes, you can set values for other metadata attributes on elements even though they're not designed for filtering or flagging. Some of the more useful attributes are importance, status, and translate.

- **Importance:** Use this attribute to specify whether content is required or optional. Setting this value is especially useful to highlight optional <step> elements in task topics.
- **Status:** Use this attribute to identify the new or changed content.
- **Translate:** Use this attribute to tell your localization team that the content contained in the element must not be translated by setting the attribute value to "no."

  If no semantically appropriate element is available to tag some of your information and you don't want that content translated, consider using a <ph> element around that content and setting the translate attribute to "no." You can also set the translate attribute on entire <topicref> elements. Setting this attribute to "no" can be helpful if you decide not to translate developer or engineering information but need to include it in a translated set of information.

TIP Instead of setting the translation attribute on all your elements, provide your localization team a list of elements that generally shouldn't be translated. For example, you might decide that content in the <apiname>, <cmdname>, and <codeph> elements must not be translated.

In addition to the importance, status, and translate attributes, you can use several other attributes for architectural or conditional processing. See the DITA specification for more information.

Topic Metadata

You can apply metadata for the entire topic in addition to applying metadata to specific DITA elements. Topic metadata is useful for writers who need to maintain DITA files and for users who want to use faceted browsing.

Topic metadata can contain information about the author, date created, and other internal, administrative information. These values can't be used for conditional processing, but topic metadata has other benefits:

- Writers can use the administrative information to maintain content over many years.
- Writers can use the metadata to locate content in a file management system.
- Users can use the metadata to filter content in a dynamic publishing or faceted browsing system.

In a DITA topic, set the metadata in the <prolog> element. Figure 8.9 shows metadata values in the <prolog> element for the Exprezzoh 9000N machine:

You can insert many different metadata elements in the <prolog> element. However, your team must decide what information goes in each of these elements or whether to use them at all. You don't need to use all the metadata available in the <prolog> element, but be consistent with the values that you do add. Create guidelines so that everyone on the team is applying metadata consistently.

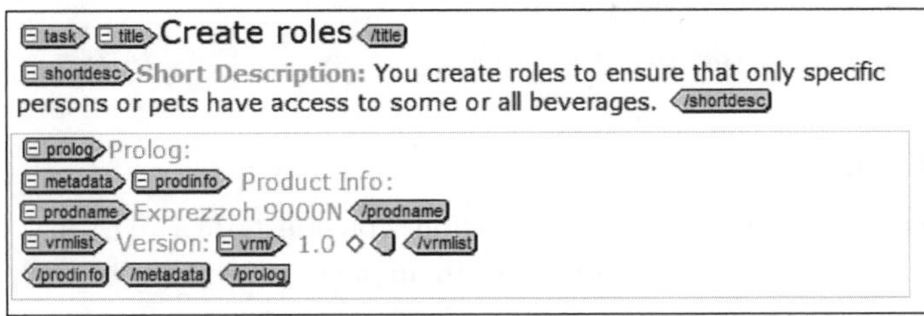

Figure 8.9 Metadata values in the <prolog> element.

Use the worksheet in Table 8.1 to help you decide which elements to use. Add your own guidelines in the last column:

Table 8.1 Guidelines for Using <prolog> Element Metadata

Reason	Types of Metadata Information Allowed in the <prolog> Element	Your Guidelines
Legal	Copyright Publisher	
Maintenance	Dates Revision information Author	

Table 8.1 Guidelines for Using <prolog> Element Metadata

Reason	Types of Metadata Information Allowed in the <prolog> Element	Your Guidelines
Search	Platform	
	Product	
	Permissions	
	Brand	
	Audience	
	Category	
Index	Keywords	
	Index terms	
	See index terms	
	See Also index terms	

DITA Map Metadata

DITA maps are the foundation for your information. You transform a DITA map to create a help system, a book, or a component of information. To improve retrievability of your information for your users and to make DITA maps easier to maintain for the writing team, you should specify metadata for DITA maps.

For example, to improve ease of maintenance for your writing team, you can specify author, copyright holder, source, and publisher, which is information that can help you maintain your files. Use the <topicmeta> element to specify this descriptive and administrative information.

You can use the <topicmeta> element in several locations in a DITA map:

- In the <map> element of a DITA map or bookmap
- In a <topicref> element
- In a <reltable> element

DITA Map Metadata

In addition to providing metadata about the entire set of information, you can include information about the topics that are included in a DITA map by using the <topicmeta> element.

There are advantages and disadvantages to applying metadata in the DITA maps or topics. Which metadata strategy you use depends on your tools and the needs of your users.

Point

Apply topic-level metadata in the DITA map.

By applying the metadata for the topic in the DITA map rather than inside the topic, the topic has more reuse potential.

Another advantage of applying topic metadata in the DITA map is that you can apply metadata to sets of topics. For example, in an installation guide, you might have a parent task topic that is written for an audience of experienced administrators. Metadata that you apply to the parent <topicref> element will apply to the child topics as well.

Counterpoint

Apply topic-level metadata in the topic.

By applying metadata in the topic, you keep the metadata values closest to the content while still allowing for reuse because setting metadata in the DITA map appends the topic metadata.

Another advantage of applying metadata in the topic is that writers can apply metadata as they write the content. This is useful especially if writers don't own the DITA maps.

Another point to consider when you are deciding on a metadata strategy is the role of content producers and content consumers in your organization. See Chapter 6, "DITA Maps and Navigation," for information about how your DITA map ownership might affect your metadata strategy.

If you decide to include topic metadata in the DITA map, apply metadata at the topic level by using the <topicmeta> element inside the <topicref> element in the DITA map. Figure 8.10 shows metadata applied to a parent <topicref> element in a DITA map.

WATCH OUT Remember that you can't use the metadata in the <topicmeta> element to filter content. You can, however, use conditional attributes to filter content when you apply metadata in the <topicref> element. For more information, see Chapter 9, "Conditional Processing."

Bookmap Metadata

The metadata in a bookmap is contained in the <bookmeta> element. You can include additional metadata items in the bookmap that you can't add in the <topicmeta> portion of a DITA map, such as the book part number, edition number, or book number. For more information about bookmap metadata, see Chapter 6 "DITA Maps and Navigation."

In Figure 8.11, metadata information is applied in the <bookmeta> element.

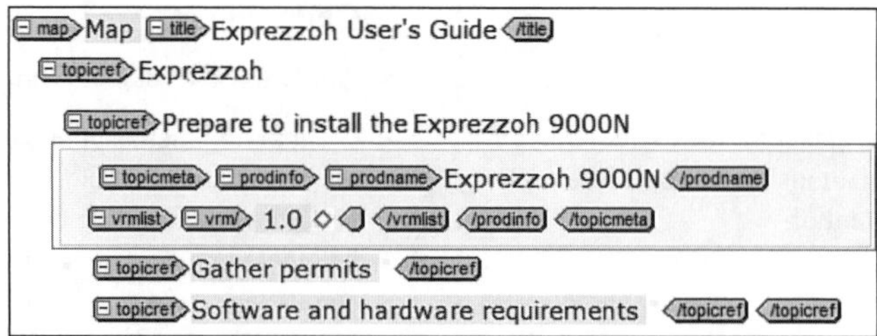

Figure 8.10 Metadata set for a <topicref > element in a DITA map.

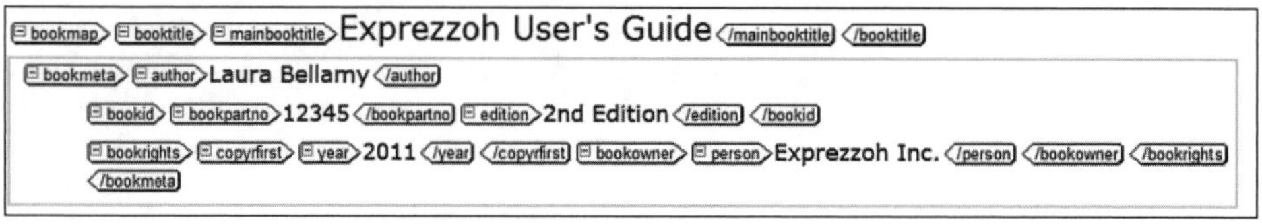

bookmap booktitle mainbooktitle Exprezzoh User's Guide /mainbooktitle /booktitle
bookmeta author Laura Bellamy /author
bookid bookpartno 12345 /bookpartno edition 2nd Edition /edition /bookid
bookrights copyrfirst year 2011 /year /copyrfirst bookowner person Exprezzoh Inc. /person /bookowner /bookrights
/bookmeta

Figure 8.11 Metadata set in the <bookmeta> element in the DITA bookmap.

Custom Metadata

If all these metadata elements and attributes still don't meet your needs, you can use the <othermeta> element to define custom metadata fields in DITA maps and topics. In the <othermeta> element, use the name and content attributes to create customized metadata entries. The name attribute specifies the property and the content attribute specifies the value.

For example, Figure 8.12 shows how you specify the region called "Europe" for this topic. You specify values for the content and name attributes for the <othermeta> element.

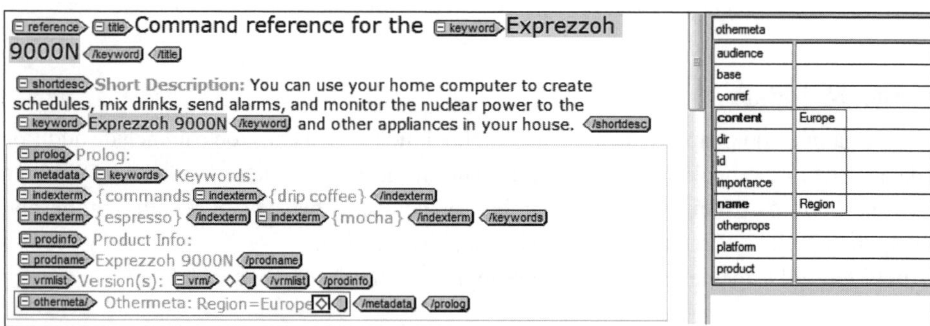

Figure 8.12 Customized metadata specified in the content and name attributes in the <othermeta> element.

TIP In HTML output, information included in the <othermeta> elements is transformed as metadata in the HTML files. Having this metadata provides a way to include search terms and keywords in your topics to improve their retrievability with search engines.

Metadata Inheritance

Before you apply metadata to elements, topics, and DITA maps, be aware that as you transform your DITA files to output, some metadata is inherited.

Standard DITA processing works from the top down, meaning that the DITA map is the highest level of the information set, and metadata is typically pushed down to other topics in the DITA map.

By default, the metadata that you apply in the DITA map will combine with or override any values that you set in the topics. For example, if you applied product and version metadata in the <topicmeta> element of a parent topic in the DITA map, the metadata values are applied to the child topics when you build output. Figure 8.13 shows the metadata from a parent <topicref> element being applied to the HTML output of both the parent and child topics.

Metadata inheritance can be a quick method for applying metadata to sets of topics in the same hierarchy. However, metadata inheritance can also lead to metadata being misapplied, which could negatively affect your output.

For example, suppose that you have a concept parent topic that describes information that's appropriate for two product guides: the regular model and the deluxe model. You set the metadata in the parent <topicref> element in the DITA map to prodname="Regular Deluxe."

However, each child topic is written for only one product. Now, when you create HTML output, the product metadata from the parent topic is inherited by the child topics. In Figure 8.14, the HTML files for the child topics have metadata values that are now incorrect. Both topics will incorrectly include metadata values for both the regular and deluxe models.

When you apply metadata, be sure that you understand how metadata is inherited:

- **Setting metadata on the DITA map:** If you set a metadata value in the DITA map's <topicmeta> element, this metadata is applied to all topics inside the DITA map.
- **Setting metadata on the parent <topicref> element:** If you set a metadata value in a parent <topicref> element, this metadata is applied to all child topics nested under the parent.
- **Setting metadata in the topic:** If you set a metadata value in the topic's <topicmeta> element in the DITA map or in the <prolog> element in the topic file, the metadata is applied to only that topic.

And just to make your head spin, you can apply metadata to any combination of DITA maps, parent topics, or other topics.

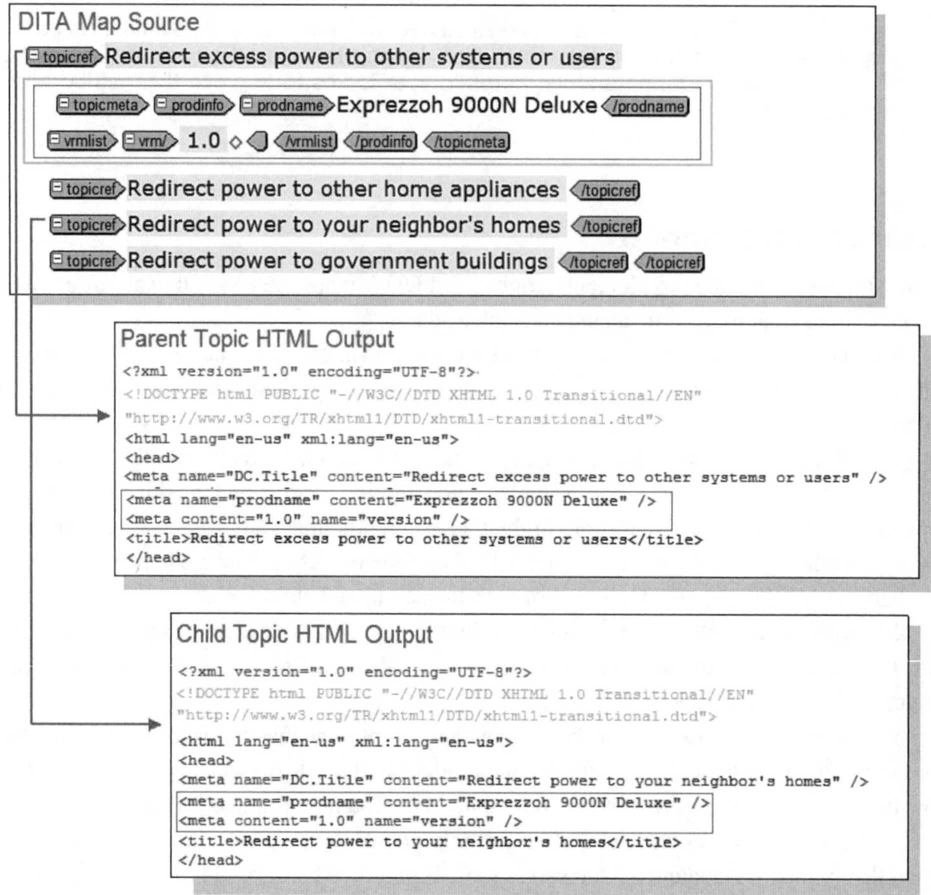

Figure 8.13 Metadata in the DITA map is inherited from parent topic to child topics.

The inheritance behavior of metadata can be confusing. Always verify that the correct metadata has been applied by checking the output.

To verify what metadata is applied, you can view the output of your HTML files and check the information in the header of your HTML files, as shown in Figure 8.15.

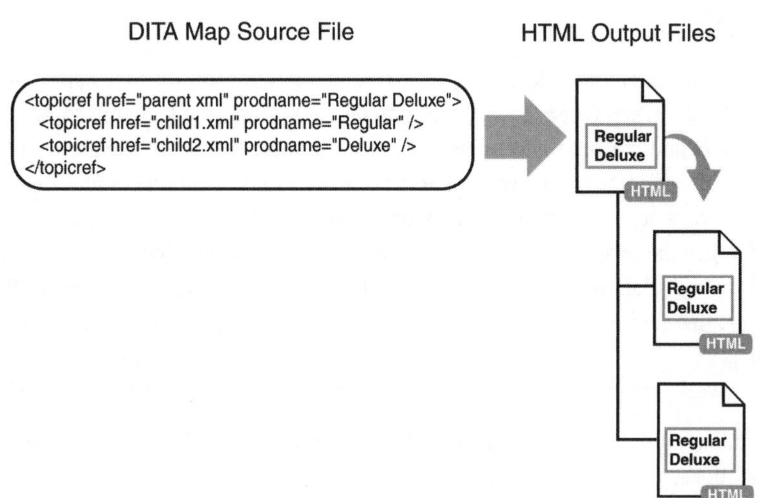

Figure 8.14 Inheritance from the DITA map and parent topics to child topics.

```
<?xml version="1.0" encoding="UTF-8"?><!DOCTYPE html PUBLIC
"-//W3C//DTD XHTML 1.0 Transitional//EN"
"http://www.w3.org/TR/xhtml1/DTD/xhtml1-transitional.dtd">
<html lang="en-us" xml:lang="en-us">
<head>
<meta content="text/html; charset=utf-8" http-equiv="Content-Type" />
<meta name="copyright" content="(C) Copyright 2005" />
<meta name="DC.rights.owner" content="(C) Copyright 2005" />
<meta content="task" name="DC.Type" />
<meta name="DC.Title" content="Connect the espresso machine to a
nuclear reactor" />
<meta name="abstract" content="Before you can use the espresso machine,
you must connect it to a local nuclear reactor." />
<meta name="description" content="Before you can use the espresso
machine, you must connect it to a local nuclear reactor." />
<meta content="reactor, connecting" name="DC.subject" />
<meta content="reactor, connecting" name="keywords" />
<meta scheme="URI" name="DC.Relation" content="ext_install.html" />
<meta scheme="URI" name="DC.Relation" content="ext_hardware.html" />
<meta scheme="URI" name="DC.Relation"
content="ext_conn_computer.html" />
<meta content="expert" name="DC.Audience.Experiencelevel" />
<meta name="prodname" content="Product B" />
<meta content="2.0" name="version" />
<meta content="Laura Bellamy" name="DC.Creator" />
<meta content="XHTML" name="DC.Format" />
<meta content="task_5F3AA1608A804F2BB6099A5976DD0A0E"
name="DC.Identifier" />
<link href="commonltr.css" type="text/css" rel="stylesheet" />
<title>Connect the espresso machine to a nuclear reactor</title>
</head>
```

Figure 8.15 Metadata information in the output of the HTML <head> element.

To Wrap Up

You've worked hard to create helpful, clear, coherent, and relevant topics. Ensure that your users can find those well-crafted topics by applying metadata to your maps, parent topics, or topic elements. Consistent and well-defined metadata not only improves the retrievability of your information, but it can also improve the way writers work with and maintain content files.

An effective metadata strategy comes only from careful planning. Assess your users and the way they'll use the product. Also, think about your product and documentation several years from now. What metadata might you need for future products, versions, or models and their corresponding documentation? Will your current metadata strategy be able to accommodate those future products?

To effectively implement a metadata strategy, you need to do the following tasks:

- Establish guidelines for all the metadata values that your team needs, including indexing guidelines.
- Decide whether to include metadata that isn't used for conditional processing, such as the <prodinfo> element.
- Decide whether to apply metadata inside the topic, on the entire topic, or in the DITA map.
- Apply metadata in your topics and DITA maps consistently. Otherwise, you'll get inconsistent and unexpected results when you build your output.
- Build your information set to verify that you're getting the correct output. Check that metadata inheritance is providing the output that you want.

Deciding how and when to apply metadata in DITA is well worth the investment.

Metadata Checklist

Guideline or Decision Point	Description
Define your users and determine how they'll use your product.	If you don't understand your users and how they use your product, you can't apply metadata effectively. Ensure that you have a well-defined metadata scheme and that you test that scheme before you implement it across your information set.
Decide which types of metadata to apply.	For most technical information, you should use: • Index entries • Conditional processing attributes • Topic metadata • DITA map metadata
Place index entries in the <prolog> element metadata.	Place index entries in the <prolog> element in most topics except for long reference topics. In long reference topics, place the index entry near the information that you want to reference.

Guideline or Decision Point	Description
Create index entries correctly.	Follow these guidelines to create effective index entries: • Ensure that every topic that has substantial content contains an index entry. • Limit the nesting levels of your index terms to secondary or tertiary. • Do not capitalize indexed terms unless they are proper nouns. • Think nouns first, verbs second. For example, make the primary entry "databases" and the secondary entry "installing." • Use plural nouns unless only singular makes sense. Be consistent. Don't create the entry "role" and then in another topic, create the entry "roles." • Avoid gerunds and imperative verbs, such as "installing" and "install" as primary entries unless no other nongerund noun is appropriate. For example, you can use "troubleshooting" as a primary entry. • Cross-post (reverse) the primary and secondary entries when possible. • Avoid inserting too many See and See Also entries. • Use the <index-sort-as> element to sort unusual terms. • Index the subjects in the topic, not just the topic title.
Decide whether to add metadata to the DITA map or to the topics.	Add metadata to the <topicmeta> in the DITA map <topicref> entries, in the <prolog> element of each DITA topic, or both. Remember to verify that your output is correct.
Decide whether to create custom metadata values.	If needed, create a custom metadata value by using the <othermeta> attribute. Set the new metadata by using the name and content attributes.
Be sure that you understand how inheritance works.	Topics inherit metadata values from maps or other topics. Test your metadata values by reviewing the HTML output of your topics.

See Chapter 9 for the conditional processing checklist.

CHAPTER 9

Conditional Processing

This is a quiz: What should you do in DITA if you have two task topics that are nearly the same except that the procedure for Task Topic A includes one extra step that is specific to a particular product model. Select the correct answer:

- **Answer 1:** Write and maintain two topics, and simply insert in the additional information in the topic that needs it.
- **Answer 2:** Write one topic that includes all the information needed for both topics and use conditional processing attributes to exclude content that you don't need.
- **Answer 3:** Take a vacation and hope someone else has the answer.

Although you were tempted to choose Answer 3, we're sure you chose Answer 2 because we know how savvy our readers are. For information that is nearly identical but has only a few differences, you don't need to create two sets of information. Instead, you create one set of topics that contains all the information needed for each product, variation, or version. Then, you apply metadata to the information in your topics and let DITA do the rest.

OK, you're right: It's not quite that simple. But by using metadata and conditional processing attributes, you can maintain only one set of source files and create variations of the output—this is the promise of single-sourcing.

For example, you might have two espresso machine models: One is the regular model, and the other is the deluxe model. The difference is that the deluxe model makes frozen coffee drinks, and so it must be connected to an ice machine. The two models share much of the same information except for a few differences in configuration and features. Thanks to conditional processing, you can write one set of topics for both models but create two user guides: one for the regular model and one for the deluxe model. The user guide for the deluxe model will include information about the frozen coffee drinks, but the regular model user guide won't.

By using a specific type of metadata called conditional processing attributes, you apply conditions. Those conditions determine which information goes into which user guide by including or excluding content when you build the output.

Figure 9.1 shows how you can use one source topic for two espresso machine models yet create different topics in the output.

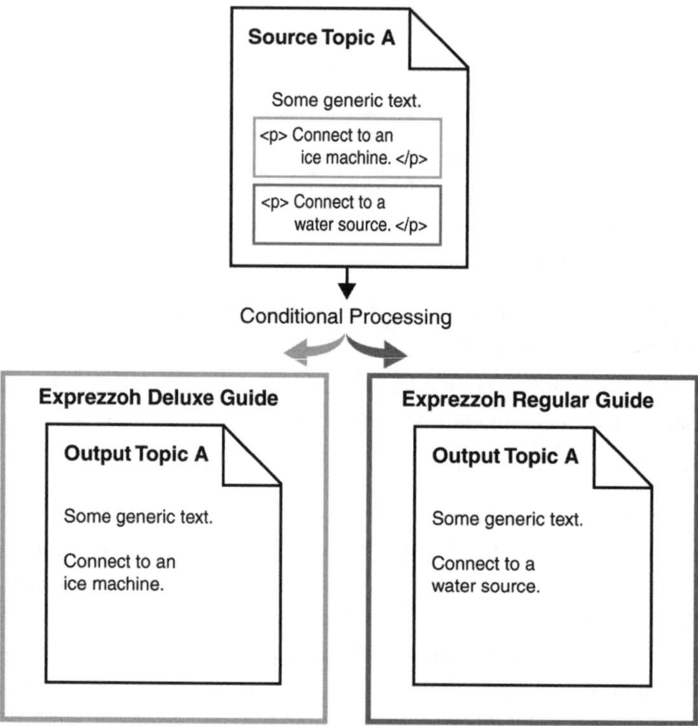

Figure 9.1 Conditional processing that can create multiple topics in output from a single source topic.

Conditional processing not only helps you create different output of a single topic, but it can also help you call out specific content by using visual aids; for example, if you want to highlight different operating systems or different product versions in your information.

Lastly, conditional processing attributes can improve how easily writers can find specific content. Writers can quickly find topics in their content management or version control systems by searching for information about the topics, such as feature numbers or product names.

To summarize, you can use conditional processing attributes to:

- Include or exclude content from output
- Visually flag content to improve usability
- Improve search and aid retrievability

For information about types of metadata other than conditional processing attributes, see Chapter 8, "Metadata."

Conditional Processing Attributes

DITA includes a standard set of conditional processing attributes that you use to process your information:

- **Audience:** Used for the intended audience of the information. For example, if your document contains information for beginners, advanced users, programmers, installers, and administrators, you can declare a unique value for each audience type.
- **Platform:** Used for operating systems or some similar environment value.
- **Product:** Used for a product or component value. For example, if you have three products or models that you provide content for, you can declare a unique value for each product.
- **Rev:** Used to record the revision level of a particular topic. You can use this attribute only to flag content. You can't use this attribute to include or exclude content.
- **Props:** A generic attribute that is used to specialize new conditional processing attributes. Don't use this attribute to filter or flag content.
- **Otherprops:** A miscellaneous attribute with no specific semantic meaning. You can use this attribute for conditional processing attribute values that don't fit in the other default attributes. So why use it? Some companies need a miscellaneous category to handle conditional processing that doesn't fit into the default attributes.

BEST PRACTICE Instead of using the otherprops attribute, consider specializing the props attribute to add new, semantically valuable attributes, such as feature, version, format, or role. *Specialization* is the process of creating a new DITA map, topic type, element, or attribute based on an existing DITA structure.

You set up conditional processing by specifying a value for the conditional processing attribute. The generic structure for applying a conditional processing attribute is:

```
<element attribute="value">
```

For example, to implement a conditional processing attribute on a paragraph element that contains content for novice users, the DITA markup is:

```
<p audience="novice">
```

Creating a Conditional Processing Scheme

Before you jump in and set values for conditional attributes, you should design a conditional processing scheme so that your team has guidelines for applying conditional processing attributes. Your scheme specifies a set of values for each conditional processing attribute so that you can create specific output. Depending on the size of your organization, you might create one scheme per product area or product set.

To design a conditional processing scheme:

1. Define your audience for the product. For example, for an espresso machine product, you might decide that some users are novices, and some are experts; novice users want to make simple coffee drinks, and experts want to be creative by mixing different beans, adding flavored syrups, and experimenting with different kinds of cream.
2. Define the operating systems, platforms, or other environments used with your product. For example, a software product might be supported on various operating systems or hardware configurations.
3. List the variations of your product by version number, model number, or other feature. For example, the espresso machine comes in three models: 9000N, 8500B, and 8000E.
4. Create unique values for the conditional attributes. For example, don't use the value "9000n" in the platform attribute to mean nuclear fusion and "9000n" in the product attribute to mean the Exprezzoh 9000N nuclear espresso machine.

A well-planned and clearly documented conditional processing scheme helps to avoid the misuse of conditions and condition sprawl. *Condition sprawl* occurs when writers create conditional values that do the same job but are defined differently. For example, one writer might set the attribute value for a platform to "windows." However, another writer might set the platform attribute to "win." And yet a third writer might use "win_only." In this case, your team must maintain three separate conditions that have the same meaning.

WATCH OUT Conditional processing values are case sensitive. For example, "Nuclear" is not the same "nuclear."

Example of a Conditional Processing Scheme

For our espresso machine user's guide, we needed to design a scheme for our conditional processing attributes. After several hours of pondering, pizza, and, of course, espresso, we've decided on the conditional processing scheme in Table 9.1.

Table 9.1 Espresso Machine Conditional Processing Scheme

Attribute	Value	Description
product	9000n	Espresso model 9000N, powered by nuclear fusion
	8500b	Espresso model 8500B, powered by battery
	8000e	Espresso model 8000E, powered by electricity (AC)
audience	expert	Experienced users of an espresso machine
	novice	New users of an espresso machines
platform	linux	Linux
	mac	Mac
	unix	UNIX
	windows	Windows

Our scheme is rather simple, but for some products the variation is much greater. For example, if you are writing an installation guide for an enterprise software product, your users might be able to install that product on five different operating systems, use three types of application servers, install it interactively or silently, and install it on one server or many servers. However, your users don't want to wade through information that doesn't apply to their environment.

To accommodate all these users, you can publish your content with these conditions:

- **Configuration 1:** Operating system 1, application server 2, silent installation, single server installation
- **Configuration 2:** Operating system 2, application server 2, silent installation, single server installation
- **Configuration 3:** Operating system 3, application server 1, interactive installation, multiple server installation
- And so on

You can see how the combinations of configurations quickly multiply. You can use conditional processing attributes to create such variation in your information, but ensure that you plan for and track how you apply those conditional processing attributes.

Remember that it takes time and effort to define a scheme and apply it accurately. When you define the scheme, you need to educate your writing team and dedicate adequate resources for writers to apply conditional processing attributes consistently throughout the content.

Applying Conditional Processing Attributes

So far, you've decided on your conditional processing scheme. Now, you're ready to apply that scheme to topics and DITA maps to:

- Include or exclude content
- Flag specific content with visual cues
- Help writers search for topics in content management systems

You can assign conditional values to include or exclude content in several places:

- In a topic to include or exclude specific elements
- In a DITA map to include or exclude entire topics
- In a relationship table to include or exclude links

Applying Conditions to Content in Topics

You can apply values to conditional attributes on most elements in your topic. Set the conditions on elements such as the <section>, <p>, <li>, <ol>, <ul>, <note>, or <codeblock> elements.

Figure 9.2 shows how the conditional value platform="unix" is applied to an <li> element.

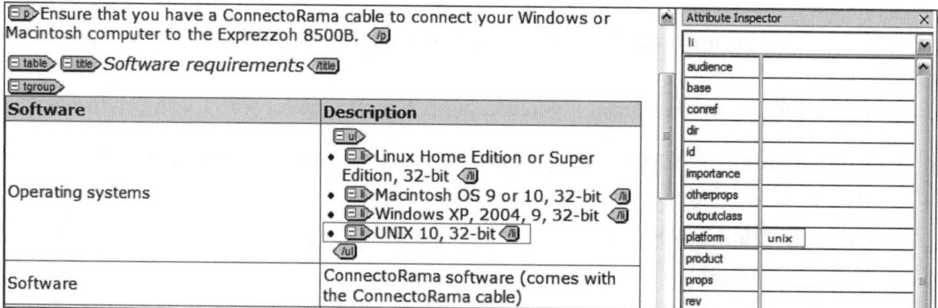

Figure 9.2 An <li> element that has the platform attribute set to "unix."

WATCH OUT We recommend that you don't apply conditions to phrase-level elements, such as <ph>, <uicontrol>, <filepath>, or <userinput> elements. Remember that you might exclude a piece of information based on the conditional processing attributes, and excluding information at the phrase level will most likely turn a sentence into nonsense.

In the case of the two espresso machines, only one is supported on UNIX®: the Exprezzoh 9000N. Therefore, when you build the output for the Exprezzoh 9000N, use processing instructions in a `ditaval` file to include elements that have the platform attribute value set to "unix."

However, when you build output for the Exprezzoh 8500B, exclude platform attributes with the value of "unix."

Figure 9.3 shows a topic that contains the following processing attribute values applied to the <p> element:

- audience="expert"
- product="9000n"

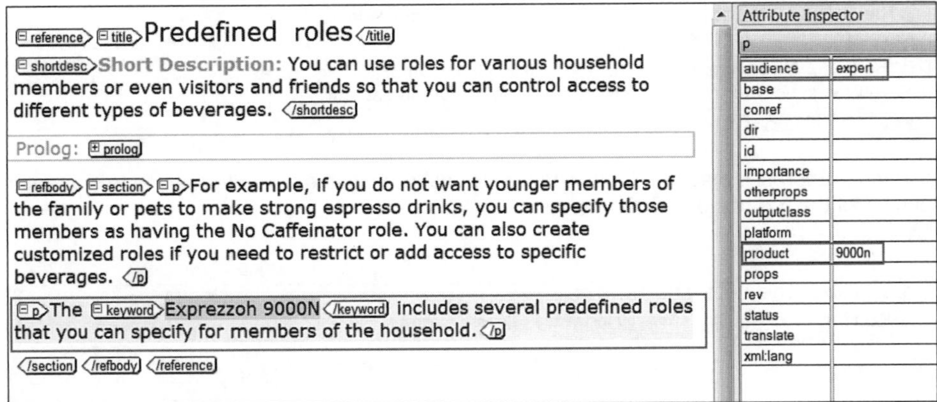

Figure 9.3 Espresso machine topic with conditional processing attribute applied to content that is specific for the Exprezzoh 9000N model.

For the nonnuclear espresso machine, which is targeted to novice users, use the following conditional processing attribute values to mark content that is specific to the Exprezzoh 8500B product, as shown in Figure 9.4:

- audience="novice"
- product="8500b"

To avoid applying a conditional value to a phrase-level element, repeat the sentence-level element.

For example, the two different espresso machines require the same step to set the time and date, but each machine has a different button name. Instead of applying a conditional value to two <uicontrol> elements, repeat the <step> element, as shown in Figure 9.5.

Then, when you create the output for your information, you see only the appropriate step, as shown in Figure 9.6.

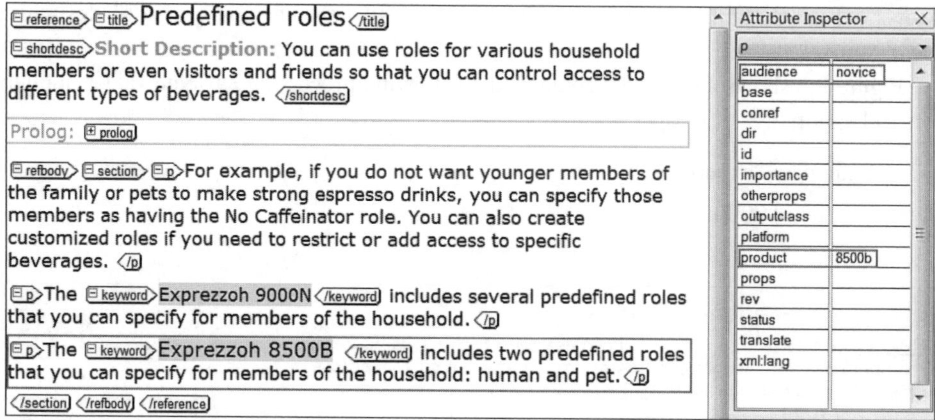

Figure 9.4 Espresso machine topic with conditional processing attribute applied to content that is specific for the Exprezzoh 8500B model.

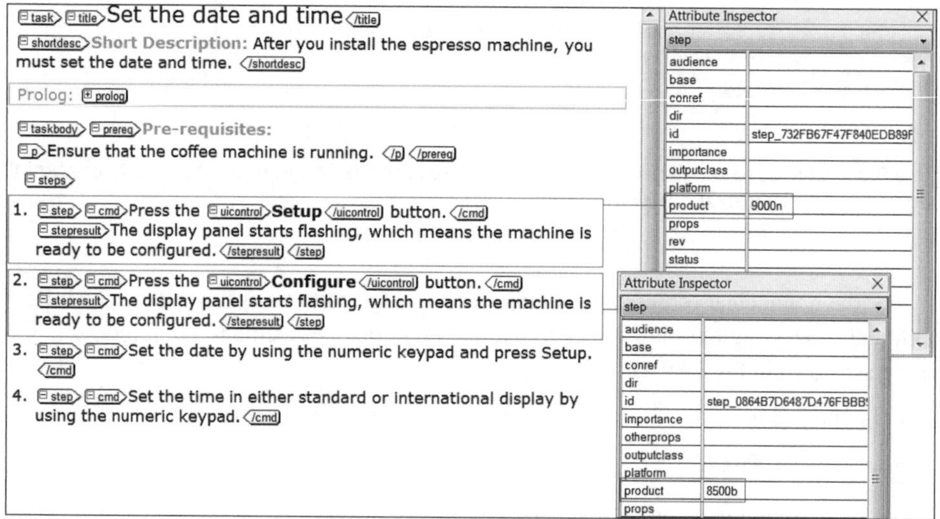

Figure 9.5 Two <step> elements that have conditional values to support two products.

If you to intend to create a guide that applies to both products, verify that the content is correct when both conditions are included. For example, if we produced an installation guide that applied to both the Exprezzoh 9000N and Exprezzoh 8500B, we need to include the information in both steps rather than exclude one step. In this case, a better option is to change the markup from two <step> elements to a <choicetable> element that describes both options in a single step, as in Figure 9.7.

Set the date and time

After you install the espresso machine, you must set the date and time.

Ensure that the coffee machine is running.

1. Press the **Setup** button. The display panel starts flashing, which means the machine is ready to be configured.
2. Set the date by using the numeric keypad and press Setup.
3. Set the time in either standard or international display by using the numeric keypad.

Figure 9.6 HTML output that shows the correct steps for the topic even though the topic contains two versions of step 1.

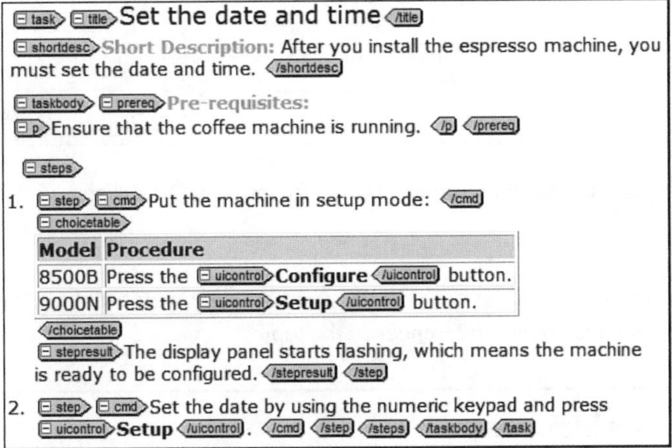

Figure 9.7 Two steps in a choice table (<choicetable> element) for two espresso machine models.

BEST PRACTICE Assign conditional values to your processing attributes as you write each topic rather than after you've written dozens or hundreds of topics. Adding these values as you write is faster than trying to hunt down topics at the end of your cycle that might be missing conditional processing attribute values.

Applying Conditions to DITA Maps and Relationship Tables

In DITA maps, you can apply conditional attribute values to the <topicref> element to exclude topics or links from your output:

```
<topicref href="exc_fusion.xml" product="9000n"> Nuclear
  fusion as a power source</topicref>
```

In Figure 9.8, we applied a condition to the <topicref> element to exclude a topic from the *Exprezzoh 8000E User's Guide*. We also applied the same condition to the <topicref> element in the relationship table to exclude a related link. If you exclude a topic from an information set, you also need to exclude any references to that topic. Otherwise, your output will contain broken links.

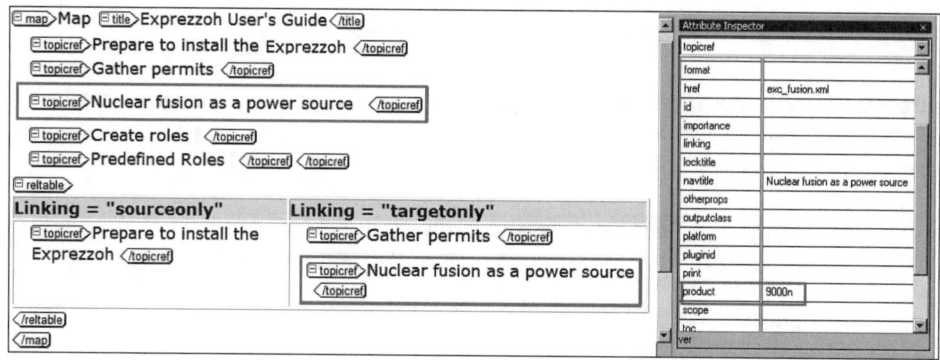

Figure 9.8 Conditions applied to the <topicref> element in a DITA map.

In the final output, the *Exprezzoh 8000E User's Guide* won't include the topic "Nuclear Fusion as a Power Source" or any references to the topic.

THE FUTURE OF METADATA

Metadata is all the rage these days, and a debate has grown in the DITA community about how metadata can help you fulfill the goals of dynamic publishing and customized content.

The current DITA standard and DITA Open Toolkit have technical limitations that restrict where you can apply metadata and how that information is processed. For example, you can control the filtering of entire topics only by applying conditional processing metadata to the <topicref> element in the DITA map. Although you can apply the same conditional processing attributes to the <task>, <concept>, or <reference> elements in a DITA file, you can't use this metadata to filter the topic.

We believe that the current implementation of DITA will evolve to support a wider and more complex range of metadata options. A future phase in publishing will likely be delivering DITA files that are dynamically consumed or displayed independently of a DITA map. For example, you might deliver your DITA XML files directly in a software product. When the user clicks a context-sensitive help button, what content is displayed depends on certain metadata in the topics, such as audience, role, product, version, and functional area. In this situation, metadata applied in a DITA map isn't available, and your tools must rely on the metadata inside the DITA topic.

Excluding and Including Content

After you apply conditional values to your topics, DITA maps, or links, you specify which conditions to include or exclude in your output by using a `ditaval` file. When you transform content to build output, the `ditaval` file specifies which content is included or excluded based on condition definitions.

In a `ditaval` file, you declare the different values in your conditional processing attributes scheme and then set those values to be included or excluded in the output. You can include only the content that you want for a specific document by including one or more attributes and values.

You can include or exclude combinations of attributes and values. Figure 9.9 shows values that a `ditaval` file might have for the espresso machine topics:

```
<?xml version="1.0" ?>
<val>
<prop att="audience" val="expert" action="include">
<prop att="audience" val="novice" action="exclude">
<prop att="platform" val="windows" action="include">
<prop att="platform" val="unix" action="include">
<prop att="platform" val="linux" action="include">
<prop att="platform" val="mac" action="include">
<prop att="product" val="9000n" action="include">
<prop att="product" val="8500b" action="exclude">
<prop att="product" val="8000e" action="exclude">
</val>
```

Figure 9.9 `ditaval` file values for the Exprezzoh machine.

Depending on your DITA authoring tool, you declare values in the `ditaval` file differently. Some authoring tools have a `ditaval` file editor, and others require that you edit the `ditaval` file in a text editor.

Consider how you want to associate `ditaval` files with sets of information. For example, you can require that each information set have a `ditaval` file, or you can reuse `ditaval` files for all information sets that have the same conditional processing schemes.

Flagging Content

You can use flagging to highlight entire topics, sentences, list items, code blocks, paragraphs, or other elements in your output. For example, in the hardware and software requirements topic for our espresso machine, you can automatically highlight Windows, Mac, and Linux® operating systems with a graphic or style by including a conditional processing value on the platform attribute. Then, you process the topics by including that conditional processing attribute value in a `ditaval` file.

Figure 9.10 shows a conditional processing value applied to the <li> element in the DITA topic to indentify content that is for Macintosh users.

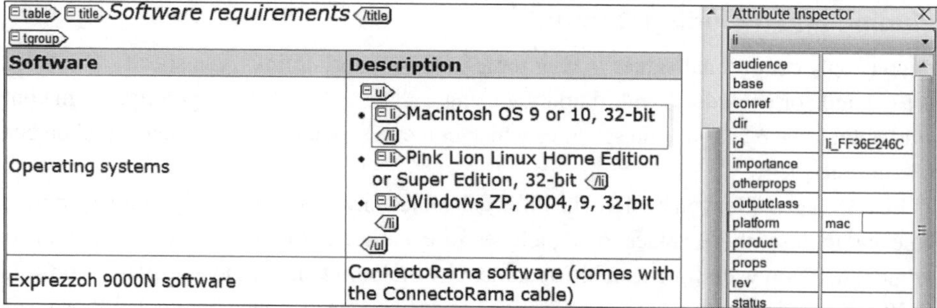

Figure 9.10 An <li> element with a platform value that flags content.

In the `ditaval` file for these topics, specify a flag on this conditional value to highlight the content with an image. Figure 9.11 shows how to use flagging by:

- Setting the action attribute of the condition to "flag" rather than "include" or "exclude"
- Using the <startflag> element to insert an image to highlight the conditional content
- Using the <alt-text> element to provide alternative text for the image

```
<?xml version="1.0" ?>
<val>
<prop att="product" val="9000n" action="include">
<prop att="product" val="8500b" action="exclude">
<prop att="platform" val="mac" action="flag">
   <startflag imageref="mac_flag.jpg">
      <alt-text>Mac content</alt-text>
   </startflag>
</prop>
<prop att="platform" val="lin" action="flag">
   <startflag imageref="lin_flag.jpg">
      <alt-text>Linux content</alt-text>
   </startflag>
</prop>
<prop att="platform" val="win" action="flag">
   <startflag imageref="win_flag.jpg">
      <alt-text>Windows content</alt-text>
   </startflag>
</prop>
</val>
```

Figure 9.11 Flagging set for Mac, Linux, and Windows operating systems in the `ditaval` file.

When you use this `ditaval` file to create output, the flagging images are added to the topic. Figure 9.12 shows the output of the topic with flagged content.

Software requirements

Ensure that you meet the following software requirements before you install the espresso machine.

Table 1. Software requirements

Software	Description
Operating systems	• ▶ Mac Macintosh OS 9 or 10, 32-bit • ▶ Linux Pink Lion Linux Home Edition or Super Edition, 32-bit • ▶ Windows Windows ZP, 2004, 9, 32-bit
Exprezzoh 9000N software	ConnectoRama software (comes with the ConnectoRama cable)

Figure 9.12 The output of conditional values used to highlight specific content.

TIP You can specify flagging to highlight content in your output in several ways. With DITA, you can underline, change text color, change background color, or wrap content in an image. The example of flagging that we demonstrate is only one option. Review the DITA specification for complete details about flagging options.

Be careful not to overuse these flags. If you find that you need to flag many items in a topic, consider creating different topics. Too many flags are distracting, and you force users to read through information that they don't need.

For example, if you're writing task information that describes how to install a client application on several operating systems, you should create one or more topics depending on the number of differences between those operating systems. Use these guidelines to help you decide when to flag differences or create separate topics:

- If you have only two or three differences in a nine-step task topic, create one topic and use flagging to highlight the differences.
- If you have more than four differences in a nine-step task topic, create one topic for each operating system.

You can devise your own guidelines about when to use flags and when to create multiple topics, but as always, use your best judgment and be consistent.

Multiple and Compound Conditions

To support more complicated conditional processing schemes, you might need to assign several conditional values to a single element to create multiple and compound conditions.

Multiple Conditions

You create *multiple* conditions when you apply two conditional processing values to the same attribute:

```
<element attribute="value1 value2">
```

For example, the following <p> element contains content that applies to both Windows and Linux platforms:

```
<p platform="windows linux">
```

Compound Conditions

You create *compound* conditions when you apply two conditional processing values to different attributes:

```
<element attribute1="value" attribute2="value">
```

For example, the following paragraph contains content that applies to the Exprezzoh 9000N product and the Windows operating system:

```
<p product="9000n" platform="windows">
```

Processing Logic for Multiple and Compound Conditions

When you create multiple and compound conditions, you must understand how processing for inclusion and exclusion works. This gets a bit technical, so grab a cup of coffee.

Content marked with *multiple* conditional values on the same attribute is included if any of the conditional processing values are set to action="include" in the `ditaval` file. However, not all conditional processing values applied to the content need to be set to action="include."

For those who prefer Boolean terms, these conditions are processed as OR operator values. That is, only one of the conditions must be true for the information to be included.

- True + True = included
- True + False = included
- False + False = excluded

As an example, the following list item has two conditional values set on the product attributes. The conditional logic in this example uses an OR operator, meaning that only one value must be true for the content to appear in the output:

```
<li platform="windows mac">
```

Table 9.2 shows when the list item appears in the output depending on the settings in the `ditaval` file.

Table 9.2 Processing Logic for Multiple Conditions on One Attribute

`ditaval` File Settings	Boolean Operator	Does the Element Appear in the Output?	Explanation
<val> <prop att="platform" val="windows" action="include"> <prop att="platform" val="mac" action="include"> </val>	OR	Yes	**Both values in the multiple condition are included:** **True:** Content marked with the "windows" value on the platform attribute is included. **True:** Content marked with the "mac" value on the platform attribute is included.
<val> <prop att="platform" val="windows" action="include"> <prop att="platform" val="mac" action="exclude"> </val>	OR	Yes	**One of the conditional values is included:** **True:** Content marked with the "windows" value on the platform attribute is included. **False:** Content marked with the "mac" value on the platform attribute is excluded.
<val> <prop att="platform" val="windows" action="exclude"> <prop att="platform" val="mac" action="include"> </val>	OR	Yes	**One of the conditional values is included:** **False:** Content marked with the "windows" value on the platform attribute is excluded. **True:** Content marked with the "mac" value on the platform attribute is included.
<val> <prop att="platform" val="windows" action="exclude"> <prop att="platform" val="mac" action="exclude"> </val>	OR	No	**Both of the conditional values are excluded.**

Content that's marked with *compound* conditional values on multiple attributes is included only if all conditional values are included. Values set on different attributes are processed as AND operator values. That is, only one processing value must be false for the content to be excluded.

- True + True = included
- True + False = excluded
- False + False = excluded

Consider the following example of a compound condition on a list item. The list item has conditional processing values set on the product and audience attributes. The conditional logic in this example uses an AND operator, meaning that all values must be true for the content to appear in the output:

```
<li platform="windows" audience="expert">
```

Table 9.3 shows when the list item appears in the output depending on the settings in the `ditaval` file.

Table 9.3 Logic for Compound Conditions on Different Attributes

`ditaval` File Settings	Boolean Operator	Does the Element Appear in the Output?	Explanation
<val> <prop att="audience" val="novice" action="include"> <prop att="audience" val="expert" action="exclude"> <prop att="platform" val="windows" action="include"> <prop att="platform" val="mac" action="exclude"> </val>	AND	No	**One of the conditional values is excluded:** **False:** Content that is marked with the "expert" value on the audience attribute is excluded. **True:** Content that is marked with the "windows" value on the platform attribute is included.

Table 9.3 Logic for Compound Conditions on Different Attributes

`ditaval` **File Settings**	**Boolean Operator**	**Does the Element Appear in the Output?**	**Explanation**
<val> <prop att=“audience” val=“novice” action=“exclude”> <prop att=“audience” val=“expert” action=“include”> <prop att=“platform” val=“windows” action=“exclude”> <prop att=“platform” val=“mac” action=“include”>	AND	No	**One of the conditional values is excluded:** **True:** Content that is marked with the “expert” value on the audience attribute is included. **False:** Content that is marked with the “windows” value on the source attribute is excluded.
<val> <prop att=“audience” val=“novice” action=“exclude”> <prop att=“audience” val=“expert” action=“include”> <prop att=“platform” val=“windows” action=“include”> <prop att=“platform” val=“mac” action=”exclude”> </val>	AND	Yes	**Both values in the compound condition are included:** **True:** Content that is marked with the “expert” value on the audience attribute is included. **True:** Content that is marked with the “windows” value on the platform attribute is included.

When you build output using the DITA Open Toolkit, the default behavior is to include content if a condition is applied to an element but isn't specified in the `ditaval` file. The DITA Open Toolkit produces the DOTJ031I informational message that indicates you have applied a conditional processing value to your content, but have not defined the condition in the `ditaval` file.

This means that if you don't specify whether to include or exclude a conditional processing value that is applied in your content in your `ditaval` file, you might expose inappropriate or confidential information to users. In other words, always specify whether to include or exclude a conditional value.

WATCH OUT Complex conditional processing can cause confusion and extra maintenance. If your conditional processing model gets too complex, consider alternative strategies. For example, use a content reference (conref) file to insert common content in a topic rather than maintain a single topic that is full of text that contains many conditions.

Identifying Applied Conditional Values

After your scheme is complete and you've started to apply conditional processing values to elements, topics, or DITA maps, you need to be sure that conditional processing attributes are correctly and consistently applied to your DITA files.

How conditional processing values are shown in the DITA source files depends on the features in your XML authoring tool. For example, you might see the conditional attribute and value directly in the element. Alternatively, you might specify a preference in your authoring tool to highlight the conditional text.

Figure 9.13 shows how to highlight conditional text in the authoring tool XMetaL.

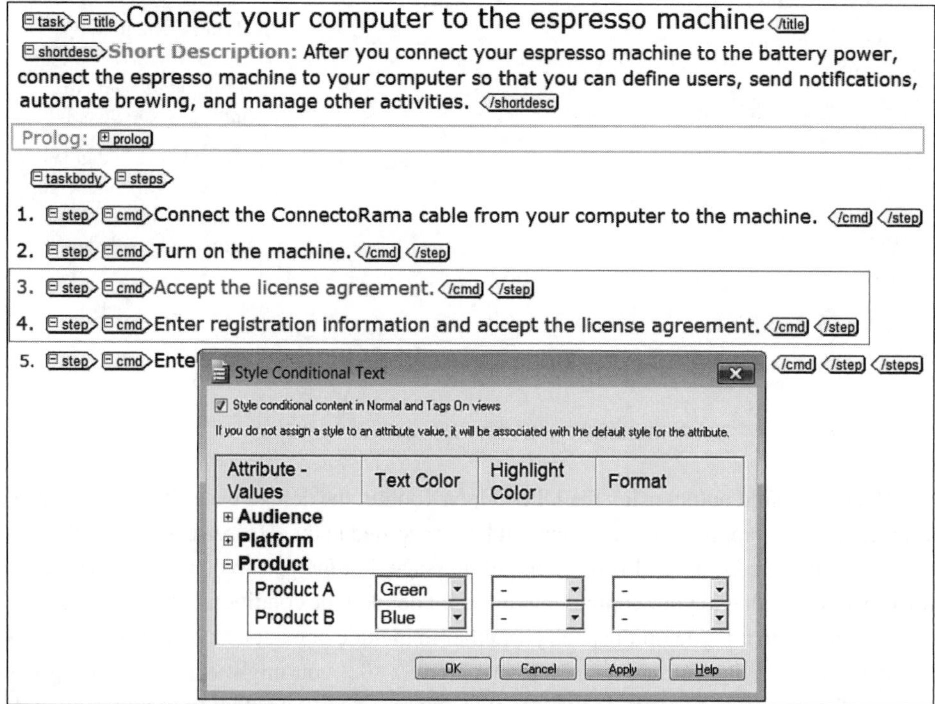

Figure 9.13 Specifying highlighting for conditional values in XMetaL so that conditional text is easier to see.

Testing Your Scheme

Before you apply your conditional processing scheme to your topic files, verify that the scheme meets your needs by testing it.

After you create the metadata scheme by defining what values will be set on what attributes—taking into account granularity and how the values will be combined—create a testing matrix as part of your test plan that shows your most popular combinations and what the expected output is.

To test your conditional processing attributes scheme, follow these steps:

1. Apply conditional processing values to topics.
2. In the `ditaval` file, exclude one value and create output of all your test topics.
3. Search the output for information that you expect to be excluded. For example, if you create a user's guide for the Exprezzoh 8500B on the Windows operating system only, you should not find any topics for the Exprezzoh 9000N, Exprezzoh 8000E, or the Macintosh, UNIX, or Linux operating systems.
4. Repeat this process for each popular combination to ensure that you're getting the expected output.

Improving Content Retrievability for the Writing Team

Investigate the features of your content management or version control system to see how conditional processing attributes can help writers to quickly locate information.

You might want to search for content that contains specific attributes and values. For example, you might want to identify all of the topics that contain audience="novice" and platform="linux." Enter a search query in your content management system that locates all topics that match the conditional processing values.

To Wrap Up

Conditional processing attributes can help you include or exclude content, flag content, or improve retrievability. By including or excluding content, you can improve your user satisfaction.

You don't need to force users to comb through information that isn't relevant to their situation. Use metadata values on conditional processing attributes to exclude information for different audiences. Your users will thank you for it.

You also don't need to maintain multiple sets of topics to accommodate slight differences in content. You can improve your productivity and reduce maintenance costs by taking advantage of conditional processing and single-sourcing of content.

You can improve the user experience by flagging, or highlighting, content by version, operating system, or other variation. However, be careful not to overuse these flags: Too much highlighting is just as bad as not enough. Rather than flag every difference in one topic, create multiple topics or exclude content for specific audiences so that users aren't overwhelmed with flagging.

And before you get too creative with specialized processing attributes, consider whether you can accommodate your users adequately with the audience, platform, and product attributes. Stick to the basic attributes if you can so that you don't unnecessarily complicate your conditional processing scheme.

Finally, before you add values for any conditional processing attributes, create and test the scheme. Take advantage of the power of DITA to create a well-tested scheme that meets the needs of your users and is simple for the writing team to apply.

Conditional Processing Checklist

Guideline or Decision Point	Description
Decide on a conditional processing scheme for your product.	Before you add values to conditional processing attributes: • Define your audiences, versions, and other distinguishing characteristics for your content. • Record the conditional processing attribute values that you'll use for your content. • Educate the team on how and when to apply the conditional processing attributes. • Ensure that everyone on the team uses the list of conditional processing attribute values.
Create `ditaval` files that specify what content to filter and flag.	Use the `ditaval file to:` • Define filtering behavior for each conditional processing attribute that should be included or excluded from the content. • Define the flagging behavior for each conditional processing attribute that should be flagged in the output. For example, do you want to insert an image or change font styles for flagged content?
Apply conditional processing attributes to content.	Apply conditional processing attributes to: • <topicref> elements in the DITA map to control the processing of topics. • Block- and sentence-level elements inside the DITA topics to control the processing of text and images. Avoid applying conditions to phrase-level elements.

Guideline or Decision Point	Description
Take advantage of the features in your XML editor to highlight elements that have conditional processing attributes set.	For example, in XMetaL, use the Style Conditional Text dialog box to define how you want conditional text highlighted.
Test your conditional processing scheme.	Testing is crucial to ensure that your output shows the expected results, especially if you use complicated conditional processing. • Decide where and when to use compound conditional values or not to use them at all. • In your conditional processing scheme, decide which content must be marked with compound conditional values and test output to ensure your scheme is working as intended.

CHAPTER 10

Content Reuse

Reusing things is not just about being a good ecological friend to the planet. In the world of DITA, reuse is about writing content once and reusing that content wherever it's needed.

You might have heard other terms that are associated with reuse, such as *single-sourcing* or *information transclusion*. Regardless of the term, reusing your content effectively in DITA can mean better content quality, consistent output, and cost savings for your company.

Benefits of Reuse

In case reuse wasn't one of your primary motivations for converting content to DITA, consider the benefits of reuse:

- **Efficiency:** Reuse improves the efficiency of writers, editors, and localization teams.
  - Writers invest the effort to create content once and reuse it many times, thereby avoiding wasted or duplicate effort.
  - Reusing content frees more time for writers to create other content.
  - When content that's reused must be updated, you reduce maintenance costs by updating only one topic instead of several.
  - You save on translation costs by translating reused content only once.
- **Consistency and accuracy:** Reuse ensures consistency and accuracy because the reused content is the same in all locations.
  - Updating only one piece of content a single time ensures that the correct changes are propagated everywhere.

   - You reduce human error by reducing the number of writers who must update the content. By using a common file to store product names and other commonly repeated text, you can better manage content that changes frequently.
- **Risk management:** Reuse can reduce the risks associated with late-breaking changes by making volatile content easier to update.

Ways to Reuse Content

DITA has an amazing set of reuse features. But before you learn about these features, here's a word of caution: You can get carried away with reuse, especially when you're getting started with DITA. Take the time to understand your motivations for reuse and develop a strategy before you implement anything.

You need to understand the use case for which you might reuse content so that you can choose the most appropriate method for reusing that content. For example, do you want to reuse product names or other common labels or objects, or do you want to reuse whole topics?

You can reuse content by:

- Reusing elements by using content references
- Reusing topics
- Reusing DITA maps
- Reusing content from non-DITA sources

You might decide, for example, to reuse product names that you insert in <keyword> elements. By reusing product names this way, you can change those product names once, and that change will be propagated to all your other topics.

Reusing Elements by Using Content References

With a *content reference*, or *conref*, you can refer to an element and use that element's content in place of the current element. You can conref topics, DITA maps, and elements so that you can reuse chunks of content from a source file in a target file. You'll most commonly use conrefs to reuse lists, steps, tables, phrases, sentences, or paragraphs in a topic.

A conref is a method of transclusion, which means that you can include a topic or part of a topic in another topic by using a reference, as shown in Figure 10.1.

For example, you might want to use a conref every time you include a table of command options in reference topics. Wouldn't you rather manage updates to the options from a single file rather than search for the options and update them in several files? With conrefs, you can change the options in just one file, and you're done.

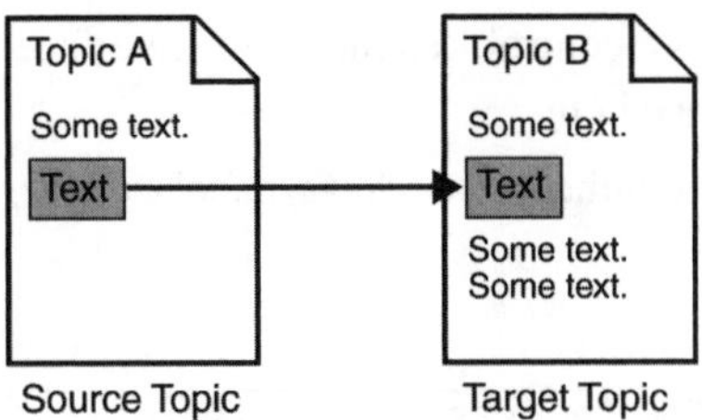

Figure 10.1 Transclusion of content.

Most technical writers have experienced the never-ending product name changes, changes to user interface labels in every release, changes to tool or object names, and changes to model numbers. And the worst part is that these changes all come at the last minute. Such is the life of a technical writer.

Using conrefs is an excellent way to reduce the risk of late-breaking or frequently changing content.

Instead of demanding that the product development team avoid making last-minute changes (which will never happen), use conrefs to reference frequently updated content from a single file. When the content changes, you modify only the file that contains the conrefs, and the update will be applied to all topics that reference that file.

For example, use conrefs for menu cascades for user interfaces. The interfaces for software products can change frequently, and having to keep up with the changes to all the <menucascade> elements in numerous task topics takes time.

Create a conref for the <menucascade> element from a designated conref file to save yourself the headache of having to manually update it in every topic for every release or iteration.

Figure 10.2 shows sections of two topics: The first topic is a designated conref file that includes all the <menucascade> elements for the Exprezzoh 9000N product; the second topic, called "Create roles," includes the conref of <menucascade> elements in the <choicetable> element in step 1.

TIP Managing frequently updated content with conrefs is also an excellent way to manage change in an agile development environment. For example, if a file path that you documented in several topics in the first sprint must be changed, you don't need to open all the files to make that change. What a time saver!

The following phrase-level elements are good candidates for conrefs. These elements often contain content that changes frequently:

- <cite> element for titles of books and information sets
- <filepath> element for file, directory, or path names
- <keyword> element for product names or model numbers

- <menucascade> and <uicontrol> elements for interface buttons, menus, check boxes, tabs, and other interface items
- <wintitle> element for names of windows, panels, pages, or dialog boxes

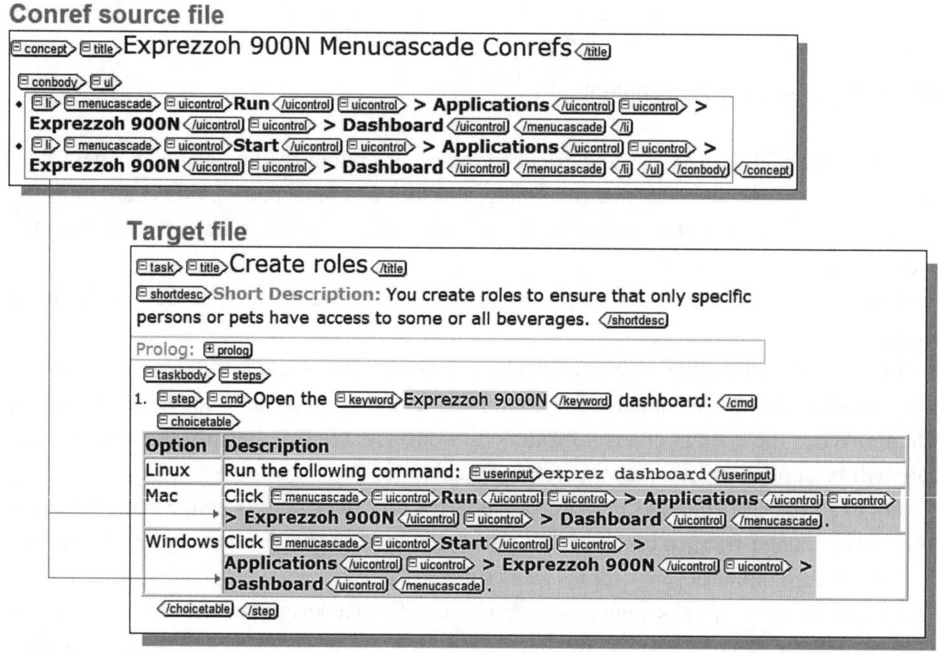

Figure 10.2 A conref of a <menucascade> element from a designated conref source file inserted in a target task topic.

Figure 10.3 shows a conref of a <filepath> element in the step of a task topic.

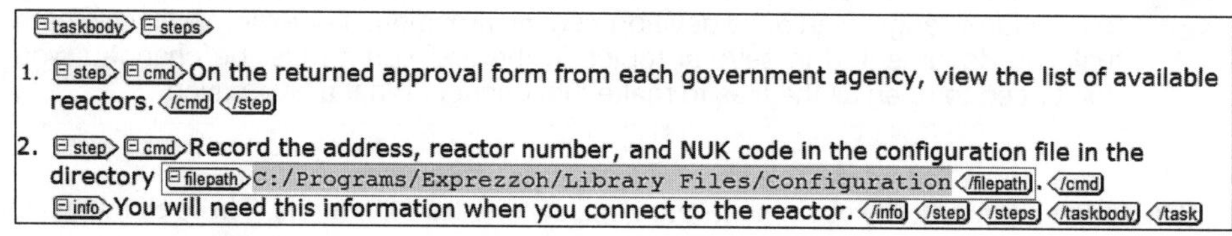

Figure 10.3 Conref of a <filepath> element.

BEST PRACTICE Be sure not to manipulate the conref content in the target topic where the conref is inserted. For example, never add an 's to make the conref text possessive, as shown in this example: `<keyword id=proper_noun> Exprezzoh 9000N</keyword>'s`. This usage can be difficult or impossible for translators to work with.

When you use conrefs for phrase-level elements, follow these guidelines:

- You don't need to reuse everything. Translation tools are pretty good at string matching. You don't need a conref for every repeated sentence. If your translation tool detects a match, it'll reuse the existing translation memory of that sentence.
- Don't use conrefs for phrase-level elements that'll complicate the translation process. Ensure that most of your phrase-level conrefs are proper nouns, such as product or model names.

Conref Attribute

Now that you're sold on the benefits of using conrefs (we hope), how do you actually implement them in DITA files? Use the conref attribute to include content from one topic in another topic. Here's the tricky part: conrefs are syntax-aware. This means that you can reuse an element and the content contained in that element only in a topic type and in a context where that element is syntactically valid. The source element for the conref must be valid in the target topic location.

In other words, say that you want to reuse a step in a task topic because that step appears in many task topics. That step might look like this:

```
<step><cmd>Log in to the dashboard as a user with the Honcho
role.</cmd></step>
```

You can reuse this group of elements and text in only the <steps> element of a task topic because only task topics accept the <step> and <cmd> elements. You can't reuse these elements in a concept or reference topic.

Whatever DITA element you use as a conref must be allowed in whatever DITA element and topic type that you want to insert the conref, for example:

- <step> element to <step> element
- <li> element to <li> element
- <p> element to <p> element
- <msgblock> element to <msgblock> element

To reuse the text from one element to another, you must set an ID on the source element so that you can establish the reference ID and reuse that element and text. Then, in the conref attribute of the target element, you specify the source topic's file name, topic ID, and the element ID as shown in Figure 10.4.

Figure 10.4 Values in a conref attribute in the target <p> element.

For example, in Topic A, if you want to reuse the text contained in a <p> element in Topic B, you first set an ID on the <p> element in Topic A. Figure 10.5 shows that the ID for the <p> element is "fusion."

The Exprezzoh 9000N requires a nuclear fusion source of power because of the extra power that the fusion process releases. The Exprezzoh 9000N cannot support less powerful forms of energy such as electricity, nuclear fission, or batteries.

dir	
id	fusion
importance	
otherprops	

Figure 10.5 The value "fusion" that's set on the id attribute for a <p> element in the source topic.

Then, in Topic B, you insert a <p> element and add a conref attribute value. The value of this attribute includes a reference to the source file and the source element that contains the content to reuse, as shown in Figure 10.6.

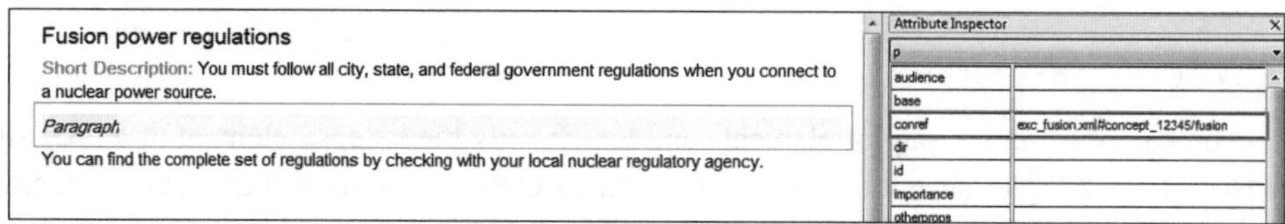

Figure 10.6 A conref inserted in the target topic "Fusion Power Regulations" as shown in XMetaL with the Tags Off view.

Fortunately, with most DITA authoring tools, you can insert conrefs by simply browsing to a target file and selecting the content instead of manually typing in the attribute value.

In the authoring tool XMetaL, the conref content appears with a gray background (as shown in Figure 10.7) to indicate that the content is referenced from another topic.

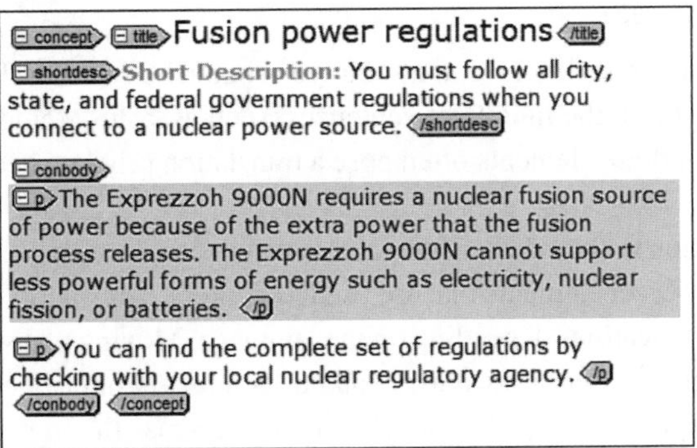

Figure 10.7 Conref attribute on a <p> element.

BEST PRACTICE If your authoring tool allows, enable automatically generated element IDs to avoid manually creating IDs for elements. Otherwise, you might need to write guidelines about how to create appropriate IDs. Manually creating IDs is time consuming. Instead, configure your XML authoring tool to automatically generate IDs on the elements that you want to reuse, as shown in Figure 10.8.

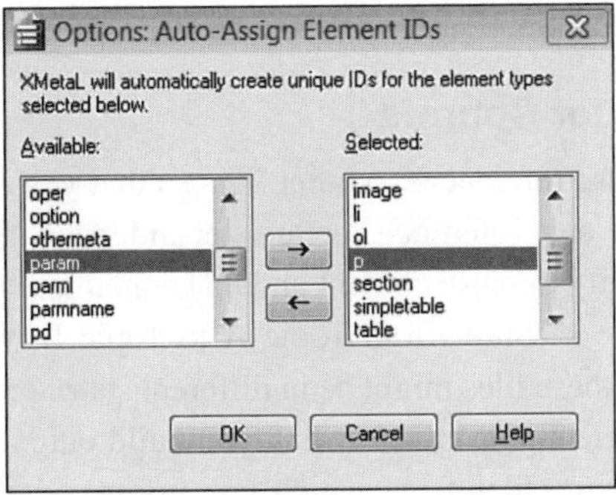

Figure 10.8 Configuring XMetaL to automatically generate element IDs.

With an automatically generated ID, the conref in the previous example would work the same way, but you don't need to bother dreaming up text for the ID. This next example shows the automatically generated ID of "p_23456":

```
<p conref="exc_fusion.xml#concept_12345/p_23456">.../p>
```

Phrase-Level Reuse

You can conref most phrase-level elements, but you should be cautious. Phrase-level elements are elements that are valid in the middle of sentences, such as <ph>, <cmdname>, and <uicontrol> elements. Conrefs of these elements often pose a translation problem because they are difficult to translate.

Some translation tools and even human translators can't properly process sentences that are broken up by phrase-level elements that are used as conrefs. Translators must move the parts of a sentence to different locations depending on the language. Moving text so that it can be translated is much more difficult if translators must contend with text in phrase-level elements that is used as a conref. Before you conref phrase-level elements, discuss the impact with your localization team to verify that translators can handle the mid-sentence references that these conrefs create. You'll need to weigh the advantages and disadvantages of using conrefs for frequently changing phrase-level text and the level of difficulty for translators that must translate that phrase-level text.

Although you might have legitimate reasons for conrefs of phrase-level elements, don't be tempted to push the limits of good markup to increase the level of reuse.

BEST PRACTICE Avoid using conrefs on content if the only way to make that content a conref is to add a phrase <ph> element around it. For example, don't use a <ph> element to conref a single sentence in the middle of a paragraph.

Designated Source Files for Conrefs

As the name suggests, conrefs use references to other files so that you can include content from those other files. Conrefs, like any reference, create dependencies between files, and these dependencies can limit your ability to reuse content and make maintaining files difficult.

For example, if you create a conref from Topic A to Topic B, you create a dependency between the two files. Although these files might be in different states or owned by different writers, both files need to be complete and available when you build output. You need to be careful about how you maintain the source files from which you insert conrefs because other writers depend on those files being correct.

If you use conrefs between active files, you risk creating spaghetti code, which means that your files have extensive and confusing dependencies, as shown in Figure 10.9.

To reduce or eliminate spaghetti code, create conrefs only from files that are designated for reuse. Identify these *designated* reuse files in such a way that writers know which files contain only conref source material.

Figure 10.10 shows conrefs created in designated conref source files. Inserting conrefs only from these designated files helps you create manageable dependencies between topics.

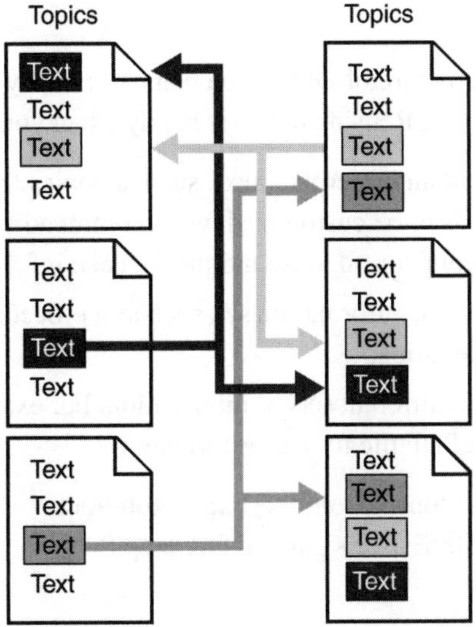

Figure 10.9 Conrefs used between active topics that create dependencies.

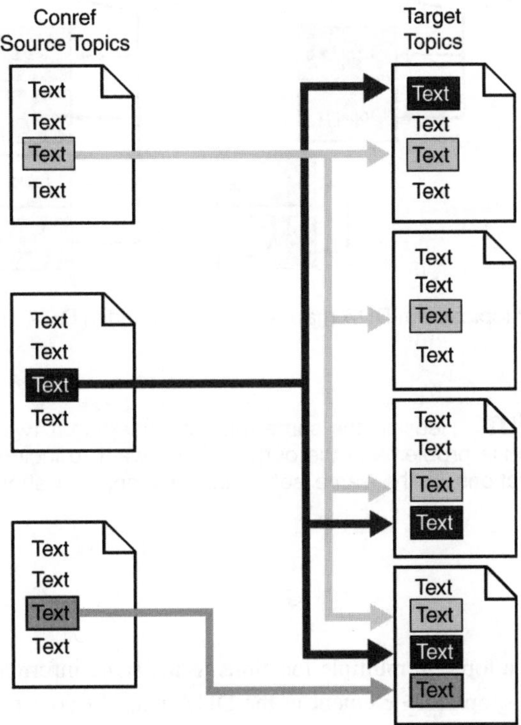

Figure 10.10 Conrefs from centralized reuse files help you to avoid creating spaghetti code.

Reusing Topics

Because of the topic-based structure of DITA, you can more easily reuse entire topics by inserting them in multiple DITA maps. Reusing topics is handy when you want to:

- Reuse topics that contain boilerplate text, such as topics that contain legal information. For content that needs to be customized, we recommend creating a DITA template that writers can then customize and make unique for each information set.
- Reuse a topic in the same information set, such as a procedure that must be done in multiple task flows or scenarios.
- Add the same topic in different sets of information. For example, add the same topic to a help system and to a PDF file for a user's guide.

In Figure 10.11, several topics from the Exprezzoh 9000N user's guide DITA map A are also used in the Exprezzoh 8500E user's guide DITA map B.

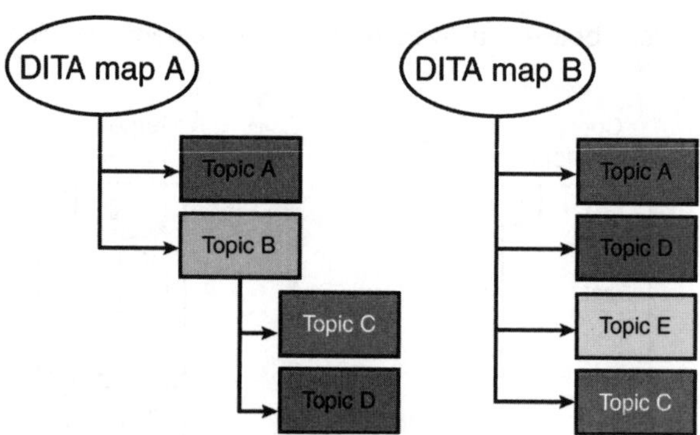

Figure 10.11 Several topics from DITA map A that are reused in DITA map B.

WATCH OUT Reusing the same topic by inserting it twice in the same DITA map can cause problems in the output. If you want to include the same topic in multiple locations in the same set of information, you should use the copy-to attribute.

Copy-to Attribute

If you want to include a topic in multiple locations in the same information set, you shouldn't simply insert the same <topicref> element in the DITA map. In your output, the topic appears twice, but all links to the topic point to the first instance of the topic.

To get around this problem, you could create a duplicate topic and use conrefs to reuse content from the original file, but that takes time, increases maintenance, and creates unnecessary files. A better solution is to use the nifty copy-to attribute to include a topic multiple times in the same DITA map or information set by reusing a single file.

You create the duplicate topic by entering a new file name value in the copy-to attribute of the <topicref> element. Because you set the copy-to attribute on a <topicref> element, you create duplicate topic references in the DITA map, but you don't actually create additional topics in the source files. For example, you can set the file name as *copy_original_filename*.dita.

BEST PRACTICE Use the <linktext> and <shortdesc> elements in the <topicmeta> element of the topic reference to provide a unique name and short description for the duplicate topic if you want to distinguish the duplicate topic in link previews.

As Figure 10.12 shows, in a DITA map, the duplicate instance of a <topicref> element uses the copy-to attribute, different link text, and a short description in the <topicmeta> element to help distinguish the reused topic from the original topic.

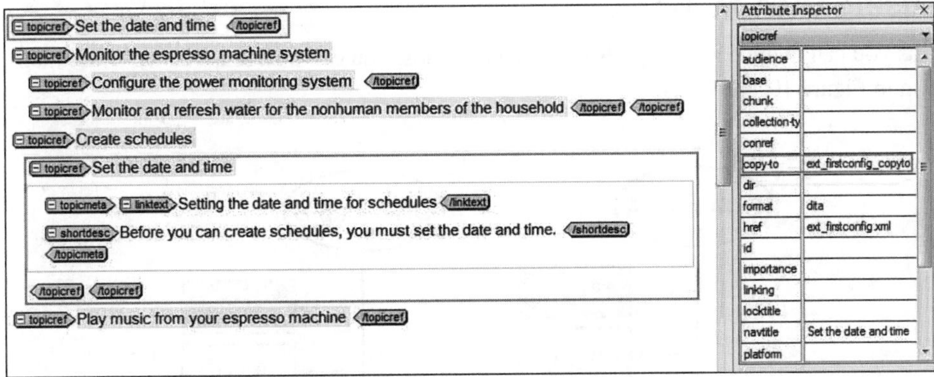

Figure 10.12 A <topicref> element that uses the copy-to attribute.

Normally, if you insert a duplicate <topicref> element in the same information set, your output shows the topic multiple times, but when you build the output unique links to the duplicate topic are not created. By using the copy-to attribute, you can create appropriate links to the reused content.

In the HTML output, although the topics are repeated, each reused topic has unique links because the files have unique file names, as shown in Figure 10.13.

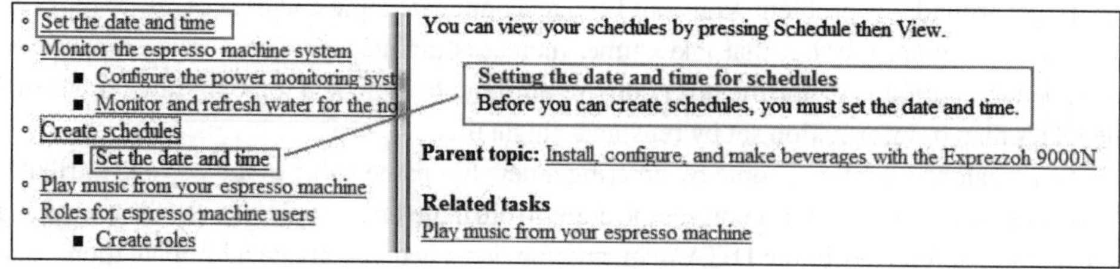

Figure 10.13 A duplicate topic with a unique link preview text.

If you want to reuse topics by including them in *different* information sets, you don't need to use the copy-to attribute. The copy-to attribute is needed only when you want to include the same topic multiple times in the same information set.

WATCH OUT You can't create duplicate instances of a submap by using the copy-to attribute. If you want to create duplicate instances of a set of topics, you must use the copy-to attribute for each topic.

Reusing DITA Maps

Just as you can include a topic in different DITA maps, you can also reuse entire DITA maps, as shown in Figure 10.14.

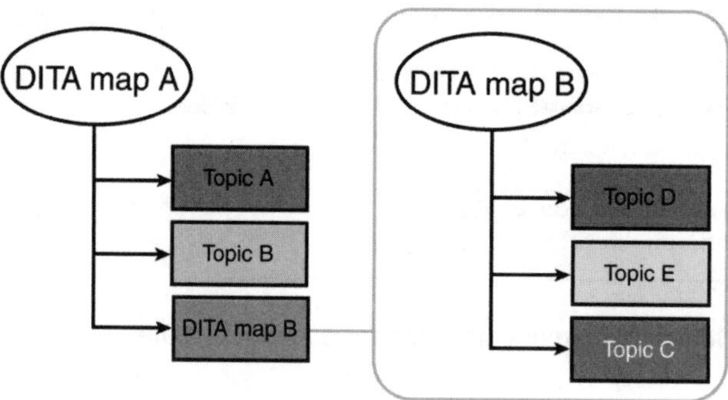

Figure 10.14 A reused DITA map nested inside another DITA map.

To reuse a DITA map, insert a <topicref> element, set the href attribute to the DITA map that you want to reuse, and set the format attribute to "ditamap."

For example, Figure 10.15 has a <topicref> element inserted in the DITA map, and its href attribute is referencing another DITA map.

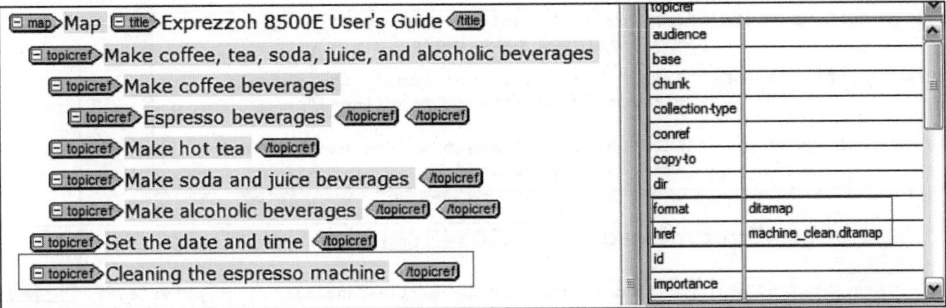

Figure 10.15 A <topicref> element with an href attribute that references another DITA map called "machine_clean.ditamap."

TIP To create reusable DITA maps, create and organize topics into submaps that document common features or user tasks. For example, even if you have one user's guide, create one DITA map for installation topics, one DITA map for security topics, and one DITA map for configuration topics. Avoid creating just one DITA map for each information set that includes hundreds of topics. See Chapter 6, "DITA Maps and Navigation," for more information about structuring DITA maps.

Reusing Content from Non-DITA Sources

Often, technical writers create content for items that are already documented in databases, product code, and other external sources. You can reuse this existing content by automatically pulling it into your DITA topics.

For example, instead of writing a programming reference guide from scratch, you might pull comments from the software code and use those comments in your DITA topics.

Because DITA is a semantic tagging language, you or even software applications can map non-DITA content to the proper DITA element and insert the content in the correct location in your DITA topics.

For example, you might be able to create and run a script that pulls messages from a product source file into DITA topics to populate your messages and codes reference book, as shown in Figure 10.16.

Writing for Reuse

As you know from the best practices for topic-based writing, well-written topics make sense in any context. (See Chapter 1 for more information.) Creating effective topic-based information provides the foundation for reusing topics.

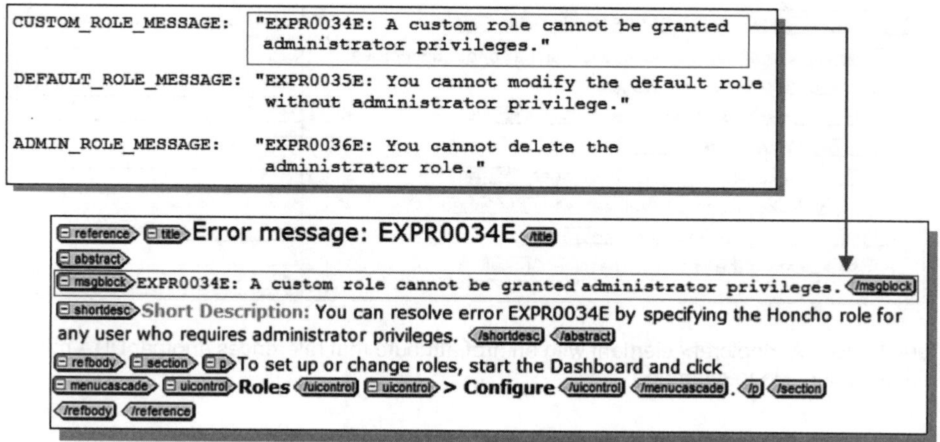

Figure 10.16 An error message that's created by pulling text from a non-DITA product source file.

To improve your ability to reuse topics, follow these writing guidelines:

- Write topics that'll make sense when taken out of context and assume that much of your information will be accessed out of context.
- Because users might read only a portion of the topic, include organizational features that make the content easy to scan and easy to carve out for reuse elsewhere. For example, help users to quickly find the relevant information in concept or reference topics by using <section> elements with titles.
- Don't use relational language. For example, avoid pointing to sentences, paragraphs, and tables when possible. Don't use statements such as "The table above describes the latest accessories." Imagine that you've reused a table from another topic but placed it after the sentence that says "above." You'll now have to rewrite your topic to remove such references.
- Write componentized content that is feature-based so that it can be reused. For example, if a particular feature such as eco-filters was added to the Deluxe model of our espresso machine, write the content so that it doesn't refer to the Deluxe model or any other features. Later, if the Regular model adopts the eco-filters feature, the topic describing these filters can be reused in both sets of information.

Deciding Which Content to Reuse

Reuse sounds great, and it can save your company time and resources, but how do you get started? When you're ready to implement reuse, you first need to identify the content to reuse. Unfortunately, there is no magical tool to identify all of your reusable content. (If we had one, we'd be selling it!)

You can do a reuse analysis to determine what content you can reuse and the best strategy for reusing it.

Step 1: Analyze Your Content

Analyze the information set or library to understand the content. When you're selecting content to reuse, consider scenarios and business goals that have common task flows.

Step 2: Identify Duplicate and Near Duplicate Content

Identify content that overlaps:

- Is the content exactly the same?
- Does the content have the same meaning but is written slightly differently?
- Is the content nearly the same except that a few pieces of content are specific to a particular product, object, or technology?

Step 3: Address the Duplication

Consider the following strategies for creating reusable topics:

- For content that is the same except written differently, rewrite the content to make it identical.
- Rewrite content to make it more generic.
- Use conditional processing in topics that have similar content so that you can reuse the topic and exclude content that is specific to a particular audience, product, or version.

Step 4: Reorganize and Rewrite for Reuse

To rewrite for reuse:

- Create reusable components. If necessary, separate content into more reusable topics and submaps.
- Consolidate duplicate content by combining common content in a single file but applying conditional processing values to different content that is meant for a specific information set.
- Use one DITA map for a specific component, area, or main user task, such as installation, security, hardware configuration, or application development.

Step 5: Implement the Reuse Strategy

Insert the reusable elements, topics, and DITA maps into your information sets. Use the copy-to and conref features and ensure that you're creating effective topic-based content.

Tips for Single-Sourcing Content Between Information Sets

When you want to share content between information sets, such as between a help system and a PDF user's guide, how do you decide what content should be reused and what content should not be reused? Consider the following tips:

- **What is the extent of reuse?** You probably don't need all your content in both the help and the manual. For example, you might have a planning guide that describes system architecture and installation prerequisites. However, you don't need to add such topics to the help system because users have already installed the product.
- **What information must users have if they can't access the help system?** For example, if users must do configuration tasks that need to be performed when the product is disabled, users need to have the information outside of the help system.
- **Which tasks require leg work?** Users need to have your documentation handy if they need to gather information from other people.
- **Who is the audience of the help and user's guide?** If the product is targeted for employees, the audience that accesses the help does not need to see the tasks that are solely for managers.

Track Your Reuse

After you implement a reuse strategy, how can you determine how much content you're reusing? Your manager would love to have some metrics about how your reuse reduces costs and how it improves efficiency and quality.

Depending on your system, you might use the following methods to determine the extent to which your information has been reused:

- Examine word-count reports in translation tools to see whether you've reduced the number of words or unique phrases.
- Count the numbers of pages that you've eliminated because of reuse.
- Use the reporting features of a content management tool to identify source and target conrefs, which can help you understand how much information is reused in conrefs.
- Determine whether reuse has increased writer productivity. Can the same writing team handle a larger project or volume of information?

To Wrap Up

DITA includes extensive tools for reusing content. Properly implemented, reuse can improve your information and can save your company money. However, reuse works only if it's used appropriately and consistently.

Carefully consider your reuse strategy before educating your writers about reuse in DITA. An overly ambitious reuse strategy can:

- Endanger topic reuse outside your library
- Confuse new writers
- Complicate translation
- Create dependencies that disrupt your componentized information model

Also, beware of reuse abuse. Don't think of conrefs as shortcuts for text that you don't want to type. Consider your motivations for reuse and agree on a strategy.

As you develop your reuse strategy, decide what types of content that you want to reuse. For example, you can reuse:

- Content in DITA elements by inserting content references (conrefs)
- Entire topics
- Entire DITA maps
- Non-DITA content

After you create a well-defined reuse strategy, establish guidelines and education for your team. Also, consider tracking metrics to show the benefits of reuse to your management team. An effective reuse strategy that's implemented consistently can save you time and resources.

Reuse Checklist

Guideline or Decision Point	Description
Define and document your reuse strategy.	Ensure that you: • Agree on a reuse strategy with your team. • Don't make your reuse strategy too complicated.
Decide which content to reuse and address potential problems with nearly identical content.	Follow these steps to identify which content to reuse: **1.** Analyze the information set or library to understand the content. **2.** Identify duplicate and near-duplicate content. **3.** Address the duplication. **4.** Reorganize and rewrite for reuse. **5.** Implement the reuse strategy.
Decide which elements are appropriate to use for conrefs.	For example, you might use the <keyword> elements to contain phrase-level content but not allow conrefs of the <ph> element.

Guideline or Decision Point	Description
When you use conrefs, establish guidelines for your team.	Remember: • Don’t try to reuse everything. • Don’t use a conref for every repeated sentence to try to reduce translation costs. Most modern translation tools can detect matches of previously translated sentences. • Use phrase-level elements such as conrefs only for proper nouns or frequently changing content. Be sure the conrefs don’t hinder translation. • Avoid using conrefs on content if the only way to make that content a conref is to add a <ph> element around it. • Create conrefs only from files that are designated for reuse to avoid creating spaghetti code.

PART III

Converting and Editing

When you choose to adopt DITA, the conversion process can be a big, but wise, investment. Before you decide to convert your content to DITA, you need a few DITA experts on your team, so be sure that your experts are knowledgeable about topic-based information, DITA maps, and linking.

Chapter 11 provides tips and guidelines to help you convert your information to DITA. Use this information to help plan and complete the conversion process.

After you convert your content to DITA or if you've been writing in DITA for years, you need to ensure that your content is of the highest quality. We recommend that you edit not only your content, but also your DITA markup, or code. Edit your DITA code so that everyone on the team is creating consistent output. And edit your content so that your users see information that's clear, consistent, complete, accurate, retrievable, and well organized.

Chapter 11

Converting Content to DITA

The decision to convert content to DITA is a big one—congratulations! You're now part of a legion of others who have discovered the benefits of DITA and topic-based writing. Now that the congratulations are over, you need to understand that converting content from an unstructured or even a structured markup language to DITA can be a daunting task. But it's well worth the effort.

The conversion to DITA requires careful planning, dedicated team work, and committed resources. You need to consider several options and strategies when you plan your conversion and agree on a solution that best suits your content and your organization.

Conversion Goals

As part of the planning process, discuss and determine your conversion goals. Avoid the "Let's just get it to DITA" strategy. Your motivations, constraints, and goals might affect your conversion strategy. Consider the following questions:

- What is the primary driver for the conversion? For example, do you need to reduce translation costs, improve customer experience, or increase the potential for content reuse?
- Are you under time constraints that require a quick conversion?
- Are you attempting a seamless transition where the output of the current content and DITA content must be identical?
- Is the conversion an opportunity to catch up with competitors? Do you need to meet or exceed industry standards?
- Do you need to modify your content to match the future corporate direction and goals?
- Is the conversion to DITA part of your adoption of topic-based writing? Do you need to do extensive reorganization of content to get it into topic form?

- To what extent will the conversion to DITA be a cultural shift for your organization? In other words, will writers have a difficult time adjusting to using semantic tagging and writing short, self-contained topics instead of traditional books?

Discuss your conversion goals with your management team, and create a strategy that addresses how these goals affect the conversion project.

Create a Pilot Team

Before you convert all of your content to DITA, you should form a pilot team and convert a small set of topics to DITA. The pilot team creates an initial conversion plan, which should include a schedule, a statement of scope, and roles and responsibilities.

Involve representatives who have the following roles on your team:

- Lead writer
- Writer
- Editor
- Information architect
- Infrastructure and tools team member
- Graphic designer
- Translation focal point
- User experience engineer
- Project manager
- Writing team manager

Involve all the appropriate stakeholders to verify that the decisions your technical experts make are in the best interest of the entire team. Most people dislike being left out and simply being told what to do. The move to DITA is going to be a big change for your organization, so use this pilot team as a way to get everyone on board with the project and excited about DITA from the start.

After you organize the pilot team members and create the initial conversion plan, select a small set of information to convert. Selecting a smaller, more manageable set of information can help you test your conversion plan, tools, and processes. This pilot conversion will also help you assess how much resource will likely be required for subsequent conversions. Don't try to convert 10,000 pages in the first go.

Conversion Process

Although you can follow different paths to convert your content to DITA, your team needs to complete the following steps regardless of whether you're doing a pilot conversion or a full production conversion:

1. Assess the state of your content.
2. Plan and schedule the conversion.
3. Prepare the content for conversion.
4. Convert the content.
5. Fix problems in the content after the conversion.
6. Reevaluate the conversion process and adjust the conversion strategy to improve the process.

Step 1. Assess the State of Your Content

What shape is your content in now? How much do you have? How much of the content needs editing and reorganizing? How much of the content is no longer needed?

Much of the discussion about converting content to DITA focuses on how the process is completed. However, the more important questions at this stage are:

- What content do you currently have?
- What content do you need to convert to DITA?
- What type of information does that content contain?
- What DITA topic types and elements do you need to support the content?

The answers to these questions will help you assess your content.

TIP If you have limited resources or little DITA knowledge to help with your conversion, consider hiring a consultant to analyze your content and make a recommendation for your conversion.

Content Analysis Worksheet

Use the worksheet in Table 11.1 to assess the content that you currently have. This information will help you to understand DITA element requirements, topic type requirements, and possible ways to reuse content.

Table 11.1 Content Analysis Worksheet

Question	Answer
What content do you intend to convert? • All the content • Only the content that's translated • Only certain kinds of information: readmes, product guides, help, knowledge base, or others • Legacy content or only content that will continue to be updated	
How much content do you have? Use a standard metric to evaluate the size of your content: • Word count • Page count • Topic count	
Do the following DITA topic types correspond well to your content, or do you need specialized topic types, such as troubleshooting or message topic types? • Concept • Task • Reference If not, describe that content.	
DITA is structured for topic-based writing. How well is your content already organized into discrete topics?	
What are the formatting or style requirements that you need to maintain or modify? Are certain pieces of text styled as bold, italic, underline, or with a specific font? How are these items styled in the output of translated content?	
Do you have inset text, reused text, or conditions that need to be maintained?	
What is your artwork sourced in and can it be output to your required formats?	

Table 11.1 Content Analysis Worksheet

Question	Answer
How do you currently translate your content, and can the translation center process DITA files?	
How do you currently build your source files to output formats such as HTML or PDF? Can DITA create those output formats?	

Step 2. Plan the Conversion

Planning for the conversion is perhaps the most difficult part of the process. You need to plan for:

- When to do the conversion
- Whether to develop the skills to do the conversion in-house or hire a vendor to do the conversion
- How to staff your team effectively
- Whether to rewrite content to topics before or after the conversion if your topics aren't already written as effective topics
- What XML standard you'll use
- How to handle various other requirements, such as graphic formats, DITA versions, conventions for file names, what topic types you need other than the three basic types, and planning for the architecture of your DITA maps

Scheduling the Conversion

The first question that teams usually ask when they decide to convert content to DITA is, "When is the best time to convert?" Hah! That's a trick question. Unfortunately, you'll never find the perfect time to convert. You'll never be given six months off from your writing and product work to focus solely on the conversion. You'll never be free from the pressure of an upcoming product release.

A better question is, "How can we plan for a successful conversion and still attend to our other responsibilities?"

The best time to convert your content is when:

- You and your team have enough resources and skills necessary to plan and complete the conversion.
- You can afford to leave the content alone for several days or several weeks. Typically, the best time to convert is between product release cycles. In addition to finding a time when writers don't need to make a lot of updates to the files, you need to account for the clean-up time that's needed after you convert your content to DITA.

- Your publishing tools are ready to support the converted content. Often teams are so focused on the source files, they forget about the output files. Remember that you must take all that converted content and create output, such as HTML or PDF files, that ship with or support your products. Your perfect DITA topics won't be much use if you can't deliver them to users. Before you begin your conversion, verify that you can publish and deliver those DITA topics in the required output formats.

Converting the Content In-House or Hiring a Vendor

A critical step in conversion planning is for you to decide who should convert your content to DITA. Should you convert the content in-house or pay another company to convert your content? We recommend that you pay a vendor. If you have the resources to work with a vendor that specializes in DITA conversions, do it.

Even though the investment seems expensive, for large conversion projects, this decision makes financial sense. When you provide conversion requirements to vendors, they create a program that runs on your content and programmatically structures the text as DITA files. Not only is the program customized to your content, but as you iterate on the conversion projects, the program will be enhanced and improved, making your future conversion projects even more efficient.

In-house DITA conversions require that some of your team members have advanced DITA skills, enough time to complete the conversion, and a knowledge of conversion tools. For companies that are just starting with DITA, the in-house DITA skills might not be developed enough to convert content as quickly or as cheaply as a vendor.

Consider the initial investment of working with a vendor against the time it would take your writers to convert the content in-house. The time that your team members spend on converting the content means they can't create or update product documentation.

The only time that an in-house manual conversion of your content to DITA is a good option is when you need to rewrite most of that content. In that case, you should probably do your own conversion. Paying a vendor to convert content that you'll extensively rewrite and reorganize would be a waste of resources.

If you do decide to convert in-house, you can use several tools to convert different file types to DITA. We've used a few of these tools during our conversion projects, and let's just say that not all tools are equal. You need to evaluate and test different tools to find the option that works best for your content.

Tips for Working with Conversion Vendors

- Ensure that you have a project manager in your company working with the conversion vendor to schedule and coordinate conversion projects.
- Hire an information architect or consultant to help you decide how content should be converted.

- Create a conversion specification that defines what the converted content must look like so that you know whether the conversion is successful.
- Look for vendor specialty. Some vendors handle only large conversion projects. Other vendors specialize in different source formats, such as HTML, FrameMaker, SGML, InDesign, or Microsoft Word.
- Carefully evaluate vendors. Shop around. Ask for sample files and customer references.
- Consider the input from your vendors. It's your project, but you hired the vendors because they're experts. Vendors have helped many companies to successfully convert content. Consider their input if they suggest ways to improve the process or your content.
- Evaluate the result of your first conversion. Consider refining the conversion process. If you don't like the result, move on to another vendor.
- If the relationship isn't working, end it. Hire another vendor rather than letting the relationship sour and the conversion project fail.

Staffing Your Conversion Team

In addition to the pilot team stakeholders, a DITA conversion requires a certain set of skills that might not exist in your team. You need to learn about the DITA standard and the tools that are required to adopt DITA. We've seen many DITA conversion projects never finish or take much longer than expected because writing teams had to learn about DITA, do a conversion, then update or reorganize topics, and do their regular writing, editing, or infrastructure work.

Realize that it takes time to learn how to use DITA. And you need more advanced skills to convert content to DITA. Avoid making your team learn, convert, and get the product out the door all in the same release cycle.

Instead of trying to staff the conversion project with people on your team, hire a consultant or vendor. If you can't hire a vendor, identify the skills that your team is missing. Which skills can you train your staff for in-house, and which do you need to outsource? Explain to your team that a conversion to DITA might require roles to change.

To do the conversion, you'll need to ensure that the conversion project team has the roles shown in Table 11.2.

Regardless of your staffing situation, understand that the adoption of DITA and the conversion effort require additional resource from the writing and editing team.

Table 11.2 Roles in a DITA Conversion Project

Role	Responsibility
Project manager	Manages the DITA conversion project
XML architect	Owns the DITA DTDs
Information architect	Defines the information model
System administrator (or tools specialist)	Owns the source control system, publishing system, and tools automation
Quality assurance engineer	Tests style sheets and transforms

Deciding on a Conversion Strategy

Assembling your conversion team, planning for the pilot project, and even hiring a conversion vendor are relatively easy tasks compared to your next big decision: deciding whether to rewrite content into proper topics before or after the conversion.

You need to choose one of these routes:

- **Reorganize content into effective topics first.** Rewrite the content in your non-DITA source to well-formed topics and then convert to DITA.
- **Convert content to DITA first.** Move the content to DITA and then rewrite and restructure content into topics as part of the cleanup effort.

In some cases, your content might already be written in good topic form. If that's the case, way to go! You're half way there. However, if your content isn't in good topic form, you have a lot of work ahead of you.

With either strategy, you might decide to convert your entire library at once or to divide the conversion effort into stages. Most companies need to use a staged approach to ensure that the time that writers can't access their content is limited.

Both routes to conversion have advantages, but what route you choose depends on:

- The state of the content before the conversion
- The resources that you have and when those resources are available
- The level of skill that your team has for creating topic-based information in DITA

Reorganize Content into Effective Topics Before the Conversion

Our editor hates it when we do bad things to words and create a term such as *topicize*, which means to rewrite content so that it's organized into effective task, concept, or reference topics. However, this term perfectly captures the task of rewriting and restructuring content to meet topic-based writing guidelines.

Rewriting and restructuring your content to topics before the DITA conversion has several advantages:

- **Your authoring tools don't matter.** You can simply use your current authoring tool to topicize your content. You and your team can do the hard work to reorganize the content by using tools that you're already familiar with.
- **Writers can work in parallel.** Writers can start topicizing content while other team members work in parallel on other conversion tasks such as assessing the content organization, mapping DITA elements to the elements or styles in your current tool, testing conversion tools, evaluating vendors, and building publishing tools. This way, writers can start early and work to prepare for the conversion.
- **Junk in is junk out.** For both in-house and vendor conversions, the conversion tools require a certain degree of consistency and quality in your content to convert the source to DITA. If your content is disorganized or inconsistent, your conversion might not go smoothly, and you'll be forced to do a lot of postconversion cleanup. Reorganizing and rewriting your content before the conversion can save you time and effort after the conversion.
- **You can remove unnecessary content.** As part of your content analysis, you should identify topics, chapters, or other pieces of information that you don't want to convert to DITA or support in future product releases. If you decide to topicize first, you can remove this content before you convert it to DITA. Don't pay to convert and clean up information that you know you don't need.
- **Source formats can be different but the output of the content is consistent.** If you're going to convert chunks of content to DITA in a phased approach, you might need to publish some content that's sourced in DITA and some that's sourced in your current format. By topicizing all the content, you ensure that all deliverables follow the same writing guidelines and have the same level of quality. Remember though that you don't need to topicize everything if you don't have the time and resources. For example, you can topicize one book at a time and then convert the topicized content to DITA later.

Convert to DITA Before Organizing Content into Topics

Converting your content to DITA now and rewriting your content as topics after the DITA conversion also has a few advantages:

- **You can work with an effective topic model.** You can write topic-based information in any source format. However, you can't always enforce good topic writing without a structure that supports topic-based writing. Because DITA provides various topic types, you'll find it easier to break up and reorganize content. The features of DITA make writing topics easier, so you might find it easier to reorganize content after it's in DITA task, concept, and reference topic files.

- **Architecture and modeling is easier when you use DITA.** You might find it difficult to reorganize content in some authoring tools. Your information architect might also want to model the content to improve the task flow and user experience. With DITA maps, you can more easily move topics to other DITA maps, submaps, or change the topic hierarchy in a DITA map.
- **DITA provides better support for tools.** When you're cleaning up converted content, you want to access, modify, and build your information regularly so that you can verify your changes in the output. Depending on your tools, access to content, and when you build the content, the publishing process might be faster with DITA content.
- **You can create consistent output.** If all your content is in DITA, it's easier to create consistent output. For example, content that is built from DITA source will have the same overall appearance for cover pages, headers, footers, and so on.

Defining Your XML Standard

Before you can convert content to DITA, you need to establish guidelines for your XML standard.

BEST PRACTICE All XML files should include an XML declaration. The declaration should specify the XML version, specify the Unicode encoding, and clearly identify the document type definitions (DTDs).

The version declaration specifies which version of XML is used and indicates to processing tools that the file is an XML document. The encoding declaration indicates to processing tools what kind of code the document uses. The most common encodings are UTF-8 and UTF-16.

Use a public ID to identify the DTDs. In an XML DOCTYPE declaration, you can use either PUBLIC or SYSTEM attributes. The public ID specifies a unique name for the DTD, whereas the system ID uses a local path that might not make sense on all systems and tools.

In Figure 11.1, the code from a DITA concept topic shows the declaration for the XML file.

```
<?xml version="1.0"?>
<!DOCTYPE concept PUBLIC "-//OASIS//DTD DITA Concept//EN" "concept.dtd">
```

Figure 11.1 The XML declaration in a DITA topic.

Establishing Graphics Formats

The DITA conversion project is an opportunity for you to improve or update the format of your graphics.

The DITA Open Toolkit supports several raster and vector graphics formats. *Raster* graphics are defined with a specific resolution and can't scale to a larger resolution without losing image quality. *Vector* graphics are drawn with vectors that enable them to maintain their sharp resolution as they scale. Depending on your content and tools, you should decide to use raster, vector, or both image formats.

In general, use the DITA conversion project as a chance to evaluate your existing graphics to determine their visual effectiveness.

Raster Graphics

The DITA Open Toolkit supports several raster graphic formats such as EPS, GIF, JPEG, JPG, and PNG. Text in these graphic formats can't be translated in the image file. Instead, the translators need to recapture the image in the appropriate language. Or if your graphic format is an output from a graphic design application such as CorelDraw (CDR), PhotoShop® (PSD), or Adobe Illustrator® (AI), the translators must have that design application to edit the source text and generate the supported output format.

TIP If your content is translated, reduce your translation costs by removing any text from the graphics and use callouts instead.

Vector Graphics

Using Scalable Vector Graphics (SVG) files as your vector graphics format has several benefits. While you're converting your text to DITA, consider converting your graphics to SVG if:

- You translate graphics that contain text.
- Your audience has a convention of using vector graphics.
- Your graphics need to scale well.
- Your graphics contain content that should to be indexed by search engines to improve retrievability.

Because SVG files are XML-based files, the text in the file is easily accessible. If you translate your content and have graphics that contain text, translators can better access and translate the text in the image file. The text in SVG files can also be crawled by search engines, which can improve the retrievability of your content.

If your company produces information for the aerospace or manufacturing industries, your audience might require highly detailed images that maintain their image quality when magnified. Vector graphics can better scale these very detailed images.

Although SVG files are best for scaling and retrievability, the format does require a modification to your publishing tools. You can use SVG files in your DITA topics without modification, but you need to modify the DITA Open Toolkit to support converting the SVG files into graphics formats that are compatible with specific output types such as PDF.

You also might need to get an additional program that allows you to view the SVG files in your XML authoring tools.

TIP Many software programs have automated tools to convert raster graphics to vector graphics. Verify whether your current tools have this conversion option. Alternatively, you can submit your files to a graphics conversion vendor.

Establishing DITA File Requirements

Before you get too far down the road to conversion, you need to decide what DITA version to use, how to name your files, how to structure your directories, and how and where to store your files.

DITA Version

DITA is a standard that's owned by the OASIS DITA Technical Committee. Each version of the standard has different functionality. We recommend that you use the latest standard and buy tools that support that version.

Your company might already have tools that help you access, write, and publish your information. Verify which DITA version the tools support. Depending on the features that you need, you might consider upgrading your tools so that you can use the latest DITA version. A tool upgrade can be an added expense to your DITA conversion process, but you should take advantage of the latest functionality that DITA has to offer.

File Names and Extensions

Create file and folder naming guidelines to ensure that your content can be published successfully and to ensure consistency throughout your documentation.

We recommend these conventions for file names:

- Use lowercase characters. Some operating systems are case sensitive. Avoid file name case conflicts by requiring that all file names are lowercase.
- Don't use spaces, special characters, or punctuation in file names. Use only hyphens and underscores to separate words in file names.
- Ensure that DITA map files use the .ditamap extension.
- For DITA topics, use either the .dita or `.xml` extension. We don't recommend that you use both extensions. Choose one extension and use it consistently.

File Naming Conventions

When you convert content from a book paradigm to topic-based architecture, you might go from a small number of chapter files that make up a book to a large number of topic files. Managing the ever increasing numbers of files can be overwhelming.

You should adopt a naming convention to make the file management easier. You can also use this conversion as an opportunity to implement a new convention.

Your version control or content management system might have strict file naming conventions. For example, some systems automatically assign a letter and number string called a global unique identifier (GUID) to each file. Rather than create a file name such as `r_fusion_regulations.xml`, your file name might be `GUID-012ABC345DEF.xml`.

If your system provides GUIDs, congratulations! You're freed from the shackles of file naming. If you don't name files by using GUIDs, you need to establish a file naming convention.

Create a file naming convention for all topics and DITA maps. Creating a file naming scheme in advance helps to limit the confusion of what files belong where or are needed. Consider the following issues to help you define file naming conventions:

- **Identify applicability.** Does your organization require that files are identified as belonging to a particular product or information set? For example, all files for one information set might have some set of characters to identify that information set. In addition, all files for that information set that are included in the same DITA map might have some set of additional characters added to the file name.

  For example, the *Exprezzoh 9000N User's Guide* might use the letters `eug` on all file names for that product. If the guide has a DITA map that's dedicated to the topics about the product installation, the name for that DITA map might be `eug_installation.ditamap`.
- **Remain open to reuse.** Are your files reused by different products and information sets? If so, don't include product or deliverable-specific information in the file name. For example, avoid using a file name like `db_security_config_guide_v2.xml`, which includes a reference to the version of the product and the specific book that it belongs to.
- **Classify your information by topic type.** Do your writers need to know the topic type based on the file name? If your version or source control system can't identify the topic type, you might want to preface each file name with a letter to indicate topic type.

  For example, a task topic might use a file name such as `t_add_users.xml`. A reference topic might use a file name such as `r_browser_requirements.xml`.
- **Optimize search engine results.** Are your file names exposed to product users in the output? Are your file names used by search engines to identify and weight your content in search results? If so, you might want to adopt a file naming convention that provides meaningful file names.

For example, this file name describes the content of the file: `exprezzoh-coffee-production.xml`.

- **Find and maintain files.** How do writers find topics in your storage system? Do you have full-text or metadata search capabilities so that writers can search for content? Or is the file name the primary way your writers find topics? If you depend on file names to understand and retrieve topics, consider using file names that clearly communicate topic content.

  For example, use a file name such as `espresso_machine_commands.xml` to describe the content of the file.

  Whatever your convention, writers need to understand how stringently they must adhere to file naming conventions. For example, if your convention is to name the file after the topic title, what happens when writers change the titles? Do they also need to update the file name? Changing file names will, of course, cause references to the old file name to be broken. What a maintenance nightmare!
- **Know your limits.** Some operating systems have file name character limits For example, the maximum path length for a Windows system is 260 characters. This character limit might seem adequate. However, be aware that you might reach that character limit if you have long file names and folder names.
- **Remember that references in DITA files might use relative paths.** If your content management or version control system does not use absolute identifiers, your ITA references might rely on relative paths. If you reference an image that's in a folder that's inside another folder, the file path can get quite long. If you establish a convention where the topic title is the file name, you could easily reach the character limit.

  For example, to make file maintenance and file naming simpler, you should avoid long file names such as:

  ```
  r_developing_automated_nuclear_fusion_reactors.xml
  ```

  If this file existed in the `upgrading` directory of the installation guide, the relative path might be very long:

  ```
  ..product_documentation/exprezzoh_installation-guide/upgrading/
  r_developing_automated_nuclear_fusion_reactors.xml
  ```

Storage System

You must store your content in a version control system. If you don't store your content in a version control system, use the DITA conversion project as an opportunity to implement a system.

A version control system provides the following benefits:

- **Security:** You know who is working on the files.
- **Revision history:** You can view previous versions of the file; you can also find information about content that was deleted or changed.

- **Ability to roll back changes:** You can roll back to a previous version of a file.
- **Automated builds:** You don't need all the files on your local file system to build output; you can start builds of output automatically when new or updated files are checked in.
- **Ability to lock content:** To eliminate change collisions, you can lock files so that no one else can change files.

  If you are implementing a version control system, consider a content management system as well. In addition to version control functionality, content management systems can also include features such as metadata management, reporting tools, and workflow.

Directory Structures

Your directory structure will have different technical requirements depending on your organization and your version control or content management system. Regardless of the tool, consider the following issues when you define your directory structure:

- If your file names do not use unique identifiers, how might references that rely on relative paths affect your directory structure? DITA uses relative paths to locate references in elements such as <xref>, <image>, and <topicref> elements. Folder paths are included in the href attribute of these elements. You should follow the same rules for folders that you do for file names: Use lowercase with no spaces or special characters.
- How do you organize topics and DITA maps that are reused in multiple information sets? Do you put reused files in a common folder? Or are reused files included with all other files?
- Do you organize content by information unit or functional area? Or do you want to break any book structures and group all files in a single directory?

Deciding What DITA Topic Types You Need

Eventually, you must determine what DITA topic types you need to structure your content. You know that the DITA standard comes with three basic topic types: concept, task, and reference. However, it also includes other topic types and several specializations. The good news is that the three basic topic types can cover the majority of your content.

For the first phase of your DITA implementation, use the default task, concept, and reference topics. Creating specialized topic types is a burden that you want to avoid when you're new to DITA.

However, if a sizable chunk of your content requires a specialized topic type, you might want to consider implementing or creating these specializations from the start.

TIP Hire a consultant to design and create the specialization for you. In the DITA community, people don't create specializations every day. Only experts understand the DITA specification and have the technical skills to create specialized topics or elements.

One Topic per File

You already know that a tenant of DITA architecture is that each topic must stand alone and have meaning independent of other topics. To enforce this best practice, establish a guideline to create only one topic per file.

WATCH OUT Don't use the composite topic type (also known as a mixed/combination topic). With the composite topic, you can mix content from two or more topic types. For example, you can combine a concept and task in the same file. This is opening the flood gates for topics that are not discrete and can't stand alone, all of which prevents reuse and reduces the retrievability of your information. If you find yourself using the composite topic type, reexamine your topic architecture.

Establishing an Architecture for Your DITA Maps

Just as you consider the conversion of content to DITA topic, you should consider how best to convert the existing hierarchy of your content to DITA map files.

As part of your conversion planning, answer these questions:

- Do you need to match the current table of contents or hierarchical structure of your topics? Or can you use the conversion project as an opportunity to reorganize your content into a different information model such as a task-oriented model?
- How will you divide content into submaps? In an automated conversion, should each chapter be converted to a submap (DITA maps inside DITA maps)? In an in-house, manual conversion, should writers be responsible for creating submaps?
- What additional DITA map elements do you want to use or avoid? For example, does your current content have book parts that would best be converted to <topicgroup> elements in a DITA map or a <part> element in a book map? Or would you rather not use <topicgroup> elements and use concept topics as containers for organizing your information instead?

BEST PRACTICE Avoid using the <topicgroup> element. This element can create empty pages in PDF output and an empty topic in HTML output. Most users are annoyed by wasted pages and hyperlinks that lead to empty pages. Instead, require that each topic in your information has meaningful content.

Handling Special Structures in Your Source Files

If you must maintain certain structures during your conversion to DITA, be sure to consider and communicate those requirements to the team.

For example, do you have variables, conditions, or text insets that you must maintain? Are there comments that record historical data that need to be left in the files? Discuss with your conversion vendor how best to convert these structures to DITA.

Step 3. Prepare the Content for Conversion

The extent to which you want to prepare your content depends on your conversion strategy. Use the checklist in Table 11.3 to evaluate which tasks you want to complete before your conversion to DITA.

Table 11.3 Conversion Preparation Worksheet

Preparation Task	Description
Follow topic-based writing guidelines.	If you decide to topicize your content before conversion, establish a checklist that writers can use to verify that content follows topic-based writing practices.
Establish topic divisions.	If your content is already in topics, deciding where to divide a chapter is much easier. If you're venturing into topic-based writing for the first time, this effort is much more difficult and time consuming. In applications such as FrameMaker, the easiest method is to break content around divisions or chapters. In HTML, break content around major sections with titles.
Identify topic types.	As you decide where to chunk your information, think about whether the specific content should be written in a task, concept, or reference topic. Be careful to convert your content to the correct DITA topic type. No authoring tool can convert content from one topic type to the other if that content is already in DITA. If you move content to the wrong type, you must manually move the content (copy and paste) to a new topic type.
Evaluate links.	Use the conversion as an opportunity to remove unnecessary cross-references and citations.
Consider nesting level limitations.	Task topics are limited to steps and substeps. If you have topics that have steps nested more than two levels, break the steps into several tasks or rewrite the task in such a way to avoid nesting steps to more than two levels. You must address this problem in the source content before the conversion to ensure that the conversion tool can process the content correctly. Otherwise, you must resolve the problem as postconversion clean up.

Table 11.3 Conversion Preparation Worksheet

Preparation Task	Description
Verify content quality.	Your non-DITA source content might use a set of tools to verify the quality of the content. Run these quality tools, such as a spelling and grammar check, before the conversion to DITA. These tools might not work with your DITA authoring tools, so this could be your last opportunity to use them to improve your content before you must adapt to a new set of quality checking tools.

After you decide which main tasks you must complete before the conversion, use the quick tips in Table 11.4 to help you identify what content belongs in what topic type.

Table 11.4 How to Identify Topic Types

Topic Type	Identifying Characteristics
Task	Although task information is often buried in conceptual or reference information in book formats, task information is the easiest to identify because it includes procedural information. Task information has the following characteristics: • Numbered lists • Describes how to do something • Verb-based titles or headings, such as: "How to start the electric lawn mower" "Configuring database security" "Prepare your home for a nuclear reactor" Be careful not to rely on the topic titles or headings alone to indicate that some information is a task. Task-like titles, such as gerund phrases, can trick you into migrating information into the wrong topic type. For example, a topic called "Understanding cat behavior" probably does not contain a procedure.
Concept	Conceptual information has the following characteristics: • Does not have numbered lists • Describes what something is or how it works • Might have bulleted lists • Noun-based titles or headings such as: "Electric lawn mowers" "Database security models" "Nuclear fusion as an alternative energy source"

Table 11.4 How to Identify Topic Types

Topic Type	Identifying Characteristics
Reference	Reference information has the following characteristics: • Often has tables and bulleted lists • Provides information meant to be found and read quickly • Describes items such as commands, objects, features, parts, or accessories • Noun-based titles or headings such as: "Electric lawn mower parts list" "Database security commands" "Average energy consumption of common appliances"

For more information about task, concept, and reference topics, see Chapters 2, 3, and 4.

Conversion Workshops

To prepare for DITA conversion, your editors, writers, and architects need to work together to complete the preparation work. Schedule conversion workshops for your team to discuss preconversion cleanup efforts and to verify that teams are correctly preparing content.

Before the content is converted to DITA, the writer, editor, and information architect should review a sample piece of the content to evaluate the preparation work. Use a checklist like the one in the Table 11.5 to track your preconversion tasks.

Table 11.5 Preconversion Tasks

Preparation Task	Status
Divide information by type: task, concept, reference.	
Rewrite deeply nested procedures.	
Move index entries near titles.	
Fix skipped headings. For example, to ensure an accurate topic hierarchy, the content shouldn't have a heading 1 followed immediately by a heading 3. Otherwise, you might get errors or unexpected results in converted DITA files.	
Remove unsupported characters or styles.	
Identify insets, comments, conditions, and variables that might be lost during conversion.	
Specify whether you want some headings and their content to be combined into one topic.	

Try converting a few pages of content to test whether you get the expected results. After this test conversion, you might identify other items to add to the checklist.

Step 4. Convert Your Source Files

Believe it or not, this is the easy part of the process. Conversion tools and vendors convert content by mapping the objects in your current information to your DITA model. Also, this is where having a project manager for the DITA conversion to manage the shipments and coordinate timing with the writers and vendors pays off. The project manager needs to ensure that the conversion runs as planned, on schedule, and within budget.

Step 5. Address Postconversion Issues

After any conversion, you'll have plenty of work to do. However, breaking up the work into phases can help your team stay on schedule and learn about DITA without being overwhelmed.

A phase is an activity that focuses on one part of the cleanup or implementation. For example, phase 1 requires that you address all the <required-cleanup> elements that are created by the conversion process.

You should complete an entire phase before moving on to the next one. These phases are designed to help writers learn about DITA gradually and allow you to ship the information to support a product release if needed even if the all the phases aren't yet done.

Phase 1: Address <required-cleanup> Elements

You must address <required-cleanup> elements that are added by the conversion tool before you can publish the content. The <required-cleanup> elements specify the items in your content that couldn't be properly converted.

TIP You can easily identify which files have <required-cleanup> elements by trying to create output of the content. The warning message DOTX039W indicates that the DITA Open Toolkit found files that contain <required-cleanup> elements.

Phase 2: Fix DITA Maps and Build Errors

Often in a conversion, the links in the original files can't be resolved in DITA. You need to check for these broken links.

You can also take this as an opportunity to delete unnecessary links. For example, if you created lists of links as a way to introduce subsections, you should delete those links. If you don't remove the links, you could get duplicate links: one set that's hardcoded and another set that's automatically created by nesting topics in a DITA map.

In this phase, work to get error-free output:

- Verify that the DITA maps have the correct hierarchy of topics. Move the topics around and change the nesting levels if necessary.
- Build output and review the build log to identify any remaining errors, warnings, or informational messages.

Phase 3: Improve Topics

In this phase, start the topicizing work if you decided to reorganize content into proper topics after you convert to DITA. In addition, use this phase to create and improve short descriptions in every topic. The most critical content to improve during this phase is the short description.

TIP Don't just move the text in the first paragraph to the <shortdesc> element. You want to carefully evaluate the paragraph before deciding that it is appropriate for the short description. For more information about writing effective short descriptions, see Chapter 5, "Short Descriptions."

During this phase, you might work closely with your information architect or editor to change the writing style of your information set from book format to a topic format. For example, remove any text that implies book type structures such as "In this chapter..." or "In this section...."

During this phase you might also improve your task flow and topic organization. We recommend the following process for further improving your converted content:

1. The information architect creates task models of the content by using an information modeling tool.
2. The writer organizes the DITA maps according to the models.
3. The writer rewrites any topics as necessary, working closely with the editor on topic writing style.

Phase 4: Check for Markup Problems and Do Code Reviews

In this phase, correct the markup in each topic to ensure that your content contains not just accurate and clear text, but also DITA markup that follows your guidelines and best practices.

During the code review phase, writers can work with their editor and information architect to ensure that the correct DITA elements are used. For more information on code reviews, see Chapter 12, "DITA Code Editing."

Phase 5: Exploit DITA

In this phase, exploit DITA so that you can reuse content, generate effective links, and improve navigation. For example, replace most inline links with either relationship table linking or by setting a collection type on a <topicref> element in the DITA map.

See Part II of this book, "Architecting Content," for more information about navigation, linking, reuse, and metadata.

Step 6. Evaluate the Conversion Process

Now that you successfully made it through converting a batch of topics, take the time to reflect on what worked well, and what didn't work well. Use this time to improve your conversion process so that the next conversion is more efficient.

To Wrap Up

Although converting to DITA can be a daunting task, creating a well-designed plan to get all your content converted ensures that your project will succeed. You must carefully assess the amount and quality of your content and then decide on a conversion process and strategy.

After the conversion is done, you still have plenty of work to do to clean up the errant markup, reorganize content, ensure that you have short descriptions, and ensure that you have no unnecessary links or duplicate links.

Most conversion projects do well if they follow these steps:

1. Do a pilot conversion.
2. Assess the state of the content.
3. Plan, plan, and plan the conversion project.
4. Prepare the content before you convert.
5. Convert the content to DITA.
6. Fix postconversion problems, reorganize content, and take advantage of DITA features.
7. Evaluate the process and project. Make adjustments as needed.
8. (Optional) Relax, have a big party, and then get ready for the next round of conversions.

You can use a checklist to monitor the progression of the conversion. This checklist is useful for keeping your conversion project on schedule and helps keep your management team informed.

Conversion Sizing Table

The following checklist is based on our conversion experience and based on the steps and phases outlined in this chapter. The tasks are weighted based on the effort compared to other tasks in your conversion. For example, if you have seven months to convert a set of information, you need 15% of that seven month period to assess the state of your content.

To get all content to DITA and to proper topics, which means using 100% of your resources, you need to plan accordingly for each phase. You can modify this table for your purposes. You can also use the table as a way to size your conversion effort.

Phase	Conversion Task	Percentage of Work	Comments
Content Assessment		15%	
Conversion Planning		10%	
Preparation of Content		35% Total	
	Follow topic-based writing guidelines	20%	If you do this work in postconversion phase 3, transfer this percentage to the phase 3 row of the table.
	Establish topic divisions	5%	
	Identify topic types	5%	
	Evaluate links	5%	
Conversion		5%	
	Conversion workshop		
	Vendor work		

Postconversion		30% Total	
	Phase 1: Address <Required-clean-up> elements.	5%	
	Phase 2: Fix DITA Maps and Build Errors . Also, build output to clean transforms.	5%	
	Phase 3: Improve topics. Topicize content, write short descriptions and index entries.	—	If the topicizing work is done after the conversion, the percentage for this phase increases.
	Phase 4: Check for markup problems and do code reviews.	10%	
	Phase 5: Exploit DITA. Insert content references (conrefs), add metadata, remove inline links, create relationship tables.	10%	
Process Evaluation		5%	

DITA Conversion Checklist

Guideline or Decision Point	Description
Assess your content.	Assess the state and quality of your content: • Decide whether it's already in effective topics. • Determine what content should move to a task, concept, or reference topic type. • Determine whether there are technical details specific to your source format; for example, text insets in FrameMaker.
Plan the conversion.	Before you plunge in and convert hundreds of topics: • Define your conversion goals. • Create a pilot team. • Determine how many people are needed for the project and create a schedule. • Decide whether to convert content to both DITA and topics or just to DITA.

Guideline or Decision Point	Description
Prepare the content to be converted.	Fix content that needs to be cleaned up before the conversion, such as stacked headings, too many nested steps, or incorrect index entries.
Convert the content.	Use a conversion vendor or in-house scripting to convert the content to DITA.
Clean up after the conversion and exploit DITA features.	Use the post conversion steps in this chapter to clean up the DITA code after the conversion and implement the different features that DITA has to offer.
Evaluate the process.	Make adjustments to sizings and schedules to make your conversion process more efficient.

CHAPTER 12

DITA Code Editing

If you want to improve the quality of your content, you should not only edit the text, but also edit the XML code, or markup, in your topics and DITA maps. You edit DITA code by doing a *code review*. The goal of the code review is to ensure that your team's use of DITA elements is consistent and appropriate.

DITA is a semantic markup language, which means you must apply DITA elements to content based on what that content is rather than how that content should appear in output. For example, elements such as <b> (bold) and <i> (italic) have no semantic value, whereas elements such as <uicontrol> or <term> describe the type of content.

That's not to say that style and highlighting are unimportant. They are important, but your markup must be consistent and correctly applied, or you're in for lots of headaches later. If you want an element to appear a certain way in output, you should change your transform and style sheet.

Applying DITA elements based on the type of content rather than the intended appearance of the output has several advantages:

- Processing systems can filter and format the content and target that content to a specific audience. If you don't apply DITA elements correctly and consistently, your content can't be properly processed.
- Sharing content is easier when writers apply DITA elements consistently. You can change output styles by using style sheets and not worry that you'll have inconsistencies.

Code reviews can help you and your team apply DITA elements correctly so that you can take advantage of the benefits of semantic tagging and topic-based architecture.

Code Reviews

A *code review* is a type of review where DITA source files are reviewed by one person or a team to identify where the markup needs to be corrected or applied. The code reviewers look for:

- **Element abuse:** Misapplied elements
- **Element neglect:** Missing or inconsistently applied elements

Code Review Benefits

DITA code reviews are an excellent way to validate your DITA markup practices and your understanding of topic-based writing guidelines.

Also, code reviews are a great way to learn how to apply DITA elements correctly. The code review is often a discovery process. Even after months or years of doing DITA code reviews, you can discover or invent new types of content that aren't addressed in your markup guidelines. You'll regularly need to:

- Decide what elements to apply to a new type of content or whether that content does not require markup.
- If no appropriate element exists for a specific type of content, create requirements for the DITA implementation at your company. For example, you might need a team to create a specialized DITA element or topic type.
- Adjust your style sheet because of new or changed style guidelines.

When you apply DITA elements consistently, you get:

- Consistent output
- Support for topic and DITA map reuse
- Simpler migration to other topic types or other elements
- Cheaper translation

Consistent Output

Inconsistencies are distracting and potentially confusing for your users. No one notices style and highlighting until something's inconsistent.

To avoid inconsistencies, use only one element for the same content even if other elements have the same output style. For example, if your style is to render the <uicontrol> and <b> elements as bold in the output, some writers might be tempted to use the <b> element to highlight button names, whereas other writers properly use the <uicontrol>.

Although the output appears correct for now, what if your team decides to change the output style of the <uicontrol> element to be monospace instead of bold? Now some button names appear in bold, and others in monospace.

Figure 12.1 shows two user interface controls with two different DITA elements: <b> and <uicontrol>.

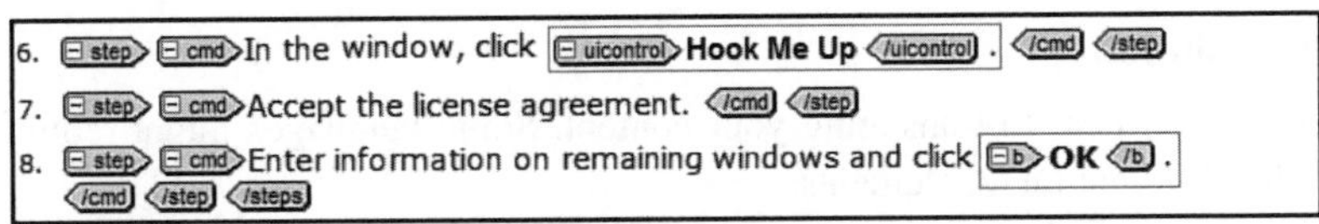

Figure 12.1 Inconsistent application of DITA elements.

When this topic is transformed to HTML, you can see the different highlighting for content that should use the same DITA element, as shown in Figure 12.2.

6. In the window, click Hook Me Up.

7. Accept the license agreement.

8. Enter information on remaining windows and click OK.

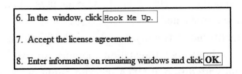

Figure 12.2 Inconsistent HTML output of content that contains inconsistent markup.

A code review can help you identify inconsistently applied elements. You can then educate your team about which elements to use and when.

Support for Topic and DITA Map Reuse

If you want to reuse either the DITA source or the output of that source with other products in your company, ensure that everyone is properly applying the DITA elements.

Suppose that you have three writing teams: Each team works on information for different espresso machine models. Some of the content for each machine is shared. For example, perhaps one team owns the tasks for how to brew espresso drinks, and another team owns the task topics for making tea and other types of drinks.

If each team uses different markup for the same types of items, the output could be inconsistent. Further, if you other teams reuse your content, you could be perpetuating inconsistencies to more product information. Such inconsistencies are difficult to correct later when you're under time and resource constraints. These inconsistencies make your content more difficult to reuse.

Simpler Migrations to Other Topic Types or Elements

When your markup is consistent, you can more easily move, or migrate, content from one topic type to another or from one element to another.

For example, suppose that you've created a specialized element called <machinecontrol> to mark up hardware button names in the documentation for your product. If some writers used the <b> element for hardware button names and other writers used the <uicontrol> element for the same hardware button names in different topics, you can't easily identify or replace the old elements with the new element without doing extra work.

Cheaper Translation

Correct markup is critical to translating your content. Some languages might require different highlighting for different DITA elements.

For example, in English, you might style the <uicontrol> element as bold, whereas in some other language, bold highlighting simply doesn't make sense, so the style sheet used for each language contains different styling than the style sheet used for English. If the source topics contain incorrect markup, the localized content could be highlighted inconsistently.

Also, many teams use specific elements to identify content that must not be translated. For example, you might decide that content in the <apiname> element must never be translated because it's code rather than plain text. However, if someone tags an API name with the <b> element, that API name might get translated.

Applying DITA elements inconsistently can also cause fuzzy matches in translation memory software. Depending on how your translation memory software works, a sentence with exactly the same content but with different DITA markup might prevent the translation software from matching the two sentences exactly, which creates a fuzzy match.

Translators rely on this matching to translate text more efficiently and accurately. Typically, the more fuzzy matches that your translators must process, the more expensive the translation service.

Identifying Code Reviewers

Code reviewers should be DITA experts, understand information architecture, and understand company style guidelines. Depending on your organization, your code review team might be a group of experts or just your most experienced DITA writer.

A code reviewer should understand quality, style, and information architecture because even though the code review is focused on the DITA markup, style and information architecture issues often arise during a review. You must occasionally decide on a strategy for new issues. For example, if an element doesn't exist for a specific type of content, you need to decide what element to apply (if any), and that decision might affect the entire or the affect the information of the entire company.

Limiting the Scope of the Review

A code review does not need to be a comprehensive review of every topic in a library. Instead, the review is typically limited to a representative set of topics. Limit how many topics you review so that you can get through the review process in an hour or two, which is a relatively small investment from the review team.

Select a set of topics of different topic types. By reviewing a variety of topic types, you ensure that everyone understands how to apply elements to each DITA topic type. We recommend that you start with at least two topics of each type: task, concept, reference, and any specialized topic types.

After you review a set of topics, those involved in the review should apply the corrections that were identified in the review to the rest of their topics. For more complex topics or if new questions or issues arise, you can do a series of code reviews on the same content.

Preparing for Code Reviews

Before you start code reviews for your teams, you need to complete these tasks:

- Establish guidelines for most or all DITA elements. If the DITA specification and language reference aren't clear about what DITA element should be applied to a specific type of content, clarify the DITA element usage in your own guidelines.
- Be sure to note what type of content you should *not* apply a DITA element to. If the appropriate element doesn't exist, it's often easier to wait for a new element to be developed (or develop it yourself) than to mark content with the wrong element and have to change it later.
- Ensure that everyone on the team uses those DITA guidelines.
- Understand how your style sheet renders the DITA elements in the output.
- Decide how to distribute and track files that are part of the code review process. Use a database, version control, or content management system to coordinate files. Avoid using email to track files and decisions.

Using Special Style Sheets for Revealing Problems in the Markup

A useful way to quickly identify markup errors is to create a cascading style sheet that can highlight common markup errors. For example, say you've wisely decided not to use the <b> element because it has no semantic value. However, new writers on your team might use this element to apply bold highlighting to content instead of applying the appropriate semantic element, such as the <uicontrol> element. You can ask these writers to build HTML output of their DITA topics and apply the special style sheet. The style sheet can highlight markup problems that might not follow your markup guidelines.

Figure 12.3 shows the code in a cascading style sheet that highlights any <b> element in a green italicized font. The style sheet in this example also highlights <userinput>, <msgph>, and <codeph> elements so that you can easily see whether those elements are applied correctly.

When you apply this style sheet and create HTML output, any instances of forbidden or potentially incorrect elements are highlighted with large, bold, and colored text, as shown in Figure 12.4.

Using this special style sheet can help writers to identify common errors before the writers submit files for the code review. If writers can correct the simpler DITA markup problems before the code review, the team can focus on more challenging problems during the code review meeting.

```
/* Start code review CSS */
strong { color:"green"; font-weight:"bold"; font-size:"large"; }
em { color: "green"'; font-style:"italic"; font-size:"large"; }
.codeph { color:"blue"; font-weight:"bold"; }
tt { color: "red"; font-weight:"bold"; text-decoration:"underline"; }
tt.msgph { color:"red" ; font-weight:"bold"; font-style:"italic"; text-decoration:"none"; }
.userinput { color "red"; font-style:"bold"; font-size:"large"; }
.sysout { color "red"; text-decoration:"underline"; font-size:"large"; }
/* End code review CSS */
```

Figure 12.3 A special cascading style sheet that reveals problems in the DITA markup.

1. On the espresso machine, press the Menu button and then select Pet Water to allow water to flow to the pet bowls.
2. From your computer, start the espresso machine dashboard.
3. From the dashboard, click **Pet Water** >> **Refresh**.
4. In the *Pet Water Refresh* page, select **Enable pet water to be refreshed**.
5. If your pet prefers the water to be swirled, select **Enable swirler**. Most cats and dogs prefer swirled water.

Figure 12.4 Example of HTML output when applying a special cascading style sheet to reveal problems.

Performing a Code Review

Code reviews are a team effort, so ensure that all the participants understand and follow the process. We recommend the following process for performing code reviews. However, you can modify this process to suit your needs.

To perform a code review:

1. The writer or DITA expert schedules the code review.

   Optional: As part of your standard editing process, you might edit the content for clarity, completeness, style, grammar, and punctuation before you do code reviews on that content.
2. The writer submits the content for the code review.
3. The code review team reviews the content.
4. The code review team and the writer discuss the findings in a meeting.
5. The writer implements findings across all topics to complete the code review.

If needed, you can schedule more code review meetings to discuss more complex issues for the same content.

Step 1: Schedule the Code Review

You can do a code review any time in the cycle, but reviewing your team's DITA markup is better done earlier rather than later. You should try to catch poor markup practices before mistakes are perpetuated throughout the content.

BEST PRACTICE Conduct at least one code review per writer each cycle. Writers often work on related content with other writers. Doing a review based on content areas, such as installation or security, rather than based on who works on the content, does not necessarily ensure that all writers have been through a code review. The goal of the code review is to verify that writers understand how to use DITA elements.

You should also schedule code reviews any time that you introduce new guidelines or topic types. A code review is an excellent way to verify that everyone understands and follows new markup guidelines. Furthermore, if you introduce new DITA topic types or templates, you might want to conduct code reviews that are specific to the new topic type.

TIP If you convert content to DITA or if you have writers who are new to DITA, schedule a code review one month after the initial DITA conversion or training. This gives new writers time to work with the DITA files and discover markup options, but it gives you the opportunity to verify markup practices before too much time has passed.

Step 2: Submit the DITA Topics for Review

Before the writers submit topics for a code review, they should ensure that the topics are valid and build error-free output. The code reviewers should not spend time debugging transform errors. If writers are having problems with transform errors, you can schedule a separate meeting or training session on debugging.

TIP Everyone's an editor, so don't make a code review about grammar, punctuation, or other content errors. Grammar, technical errors, or even bad writing can be distracting in a code review when you should focus on only the DITA markup. To prevent an editorial free-for-all, ask writers to request an edit of the content before they get a code review.

Step 3: Review the DITA Markup

The writer sends the DITA files to the DITA expert to review and identify markup that must be corrected. To mark changes, the DITA expert can use one or more of the following options:

- Add comments in the DITA <draft-comment> or XML comment elements where changes are required.
- If available, use a tool to track changes in the XML editor.
- Make changes directly to the topic, but ensure that everyone sees the old and the new version of the file.

After the DITA expert reviews the files, he or she sends the files to the other code review team members to verify the suggestions and to decide how to handle style issues uncovered during the code review. Sending the DITA files to one reviewer at a time can save time by preventing reviewers from making the same comments twice.

After spending hundreds of hours doing code reviews, we've noticed many recurring problems with DITA markup, such as problems with steps in task topics or using definition lists to create extra headings in concept topics.

Common Problems Found in Task Topics

In task topics, you often find the following markup problems:

Problem: Using an ordered list (<ol> element) instead of the <steps> element

Incorrect

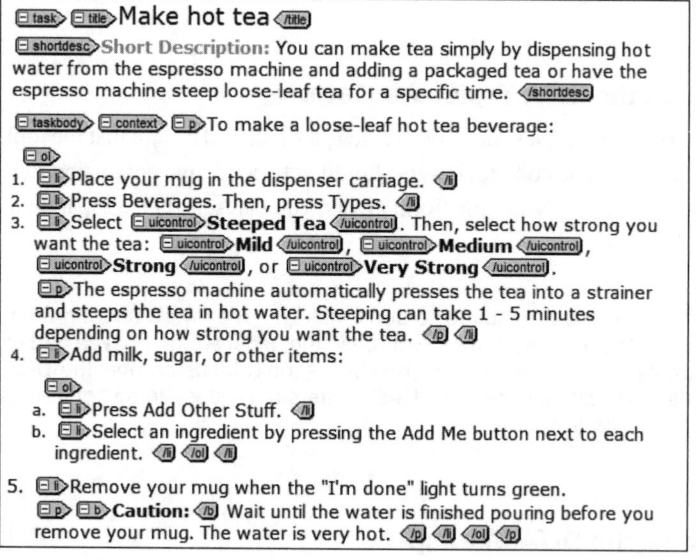

Figure 12.5 A task topic with an inappropriate ordered list in the <context> element.

Correct

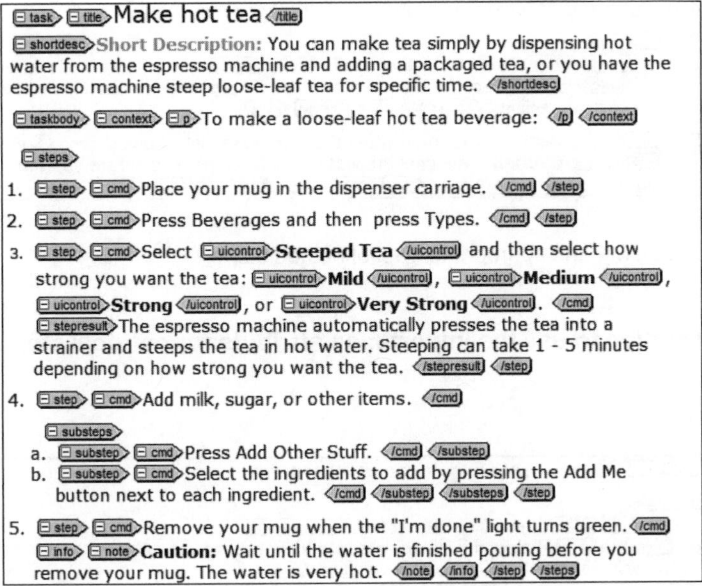

Figure 12.6 A task topic that correctly uses the <steps> element for numbered steps.

Problem: Creating too many levels of nested steps; for example, using the <info> and <ol> elements in a <substep> element to add a third level of ordered steps

Incorrect

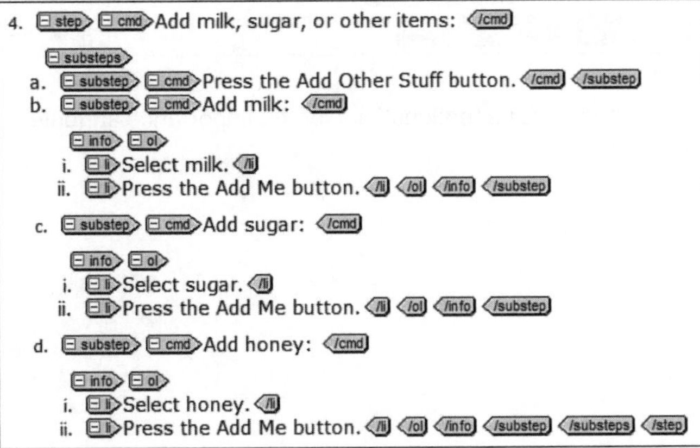

Figure 12.7 A task topic with too many levels of nested substeps.

Correct

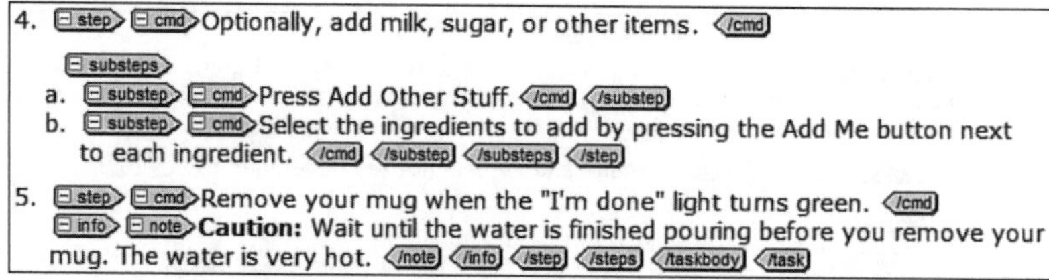

Figure 12.8 A task with the appropriate level of nested steps.

Problem: Forgetting to use the importance attribute for optional steps

Incorrect

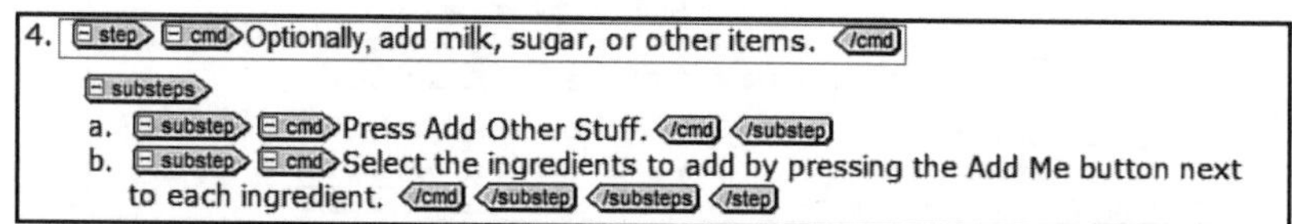

Figure 12.9 Using the term "Optionally" to indicate an optional step instead of setting a value for the importance attribute.

Correct

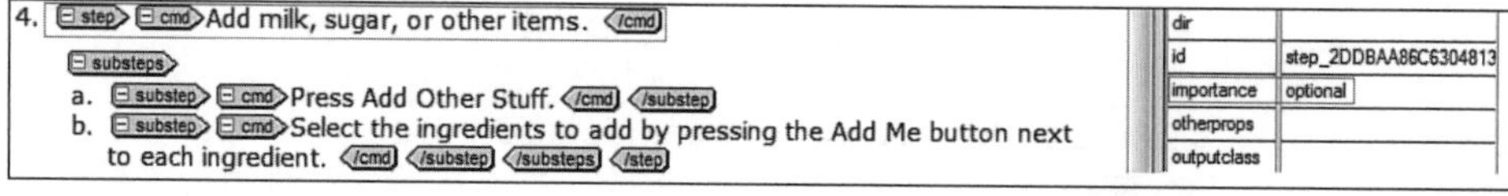

Figure 12.10 Step 4 with the value "optional" set for the importance attribute, which creates the label "Optional" in the output.

Problem: Using the <cmdname> element instead of the <userinput> element in task steps where the user must enter a command

Use the <userinput> element to show what the user must type or enter in a computer interface to complete a step. Use this element if that entered data is short or takes up no more than one line.

In software documentation, use the <cmdname> element for command names when you're describing the command and not showing users how to enter the command.

Incorrect

1. step cmd To start the installation program from a command line, enter cmdname ~install exprezzoh~ /cmdname. /cmd
info The program starts checking for prerequisite software automatically. /info /step

Figure 12.11 A step that incorrectly uses the <cmdname> element instead of the <userinput> element.

Correct

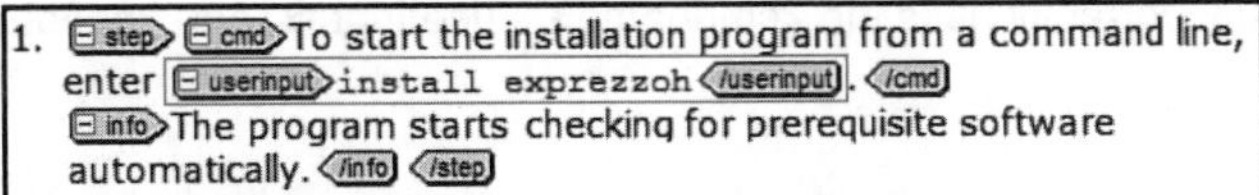
1. step cmd To start the installation program from a command line,
enter userinput install exprezzoh /userinput. /cmd
info The program starts checking for prerequisite software
automatically. /info /step

Figure 12.12 A step that correctly uses the <userinput> element.

TIP For commands, use the <codeblock> element for data that is two lines or more, especially if the user is expected to press Enter after each line. You can also use the <codeblock> element to show other types of code or data.

Figure 12.13 shows the correct examples for the <cmdname>, <userinput>, and <codeblock> elements. The <context> element contains the command name of exprez in the <cmdname> element; step 1 includes a short command in the <userinput> element, and step 2 contains a long command in the <codeblock> element:

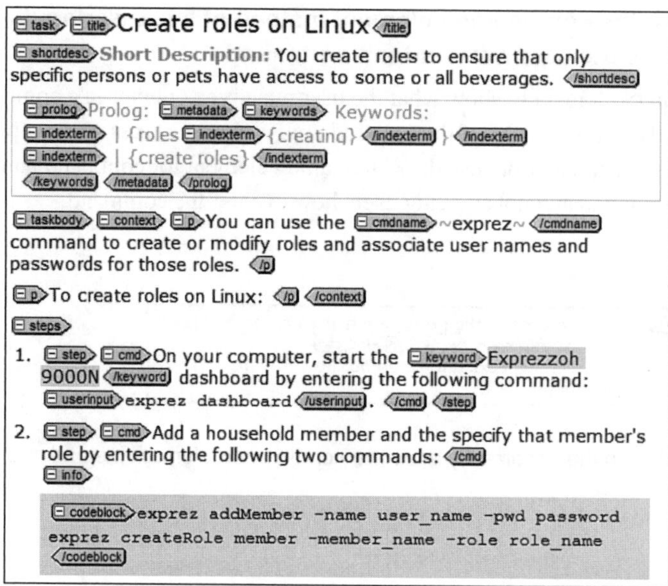

Figure 12.13 A task topic that correctly uses the <cmdname>, <userinput>, and <codeblock> elements.

Problem: Forgetting to mark window, page, or panel names in software user interfaces with the <wintitle> element

Incorrect

4. step cmd In the Pet Water Refresh page, select uicontrol **Enable pet water to be refreshed** /uicontrol . /cmd /step

Figure 12.14 A step that's missing the <wintitle> element around the page called "Pet Water Refresh."

Correct

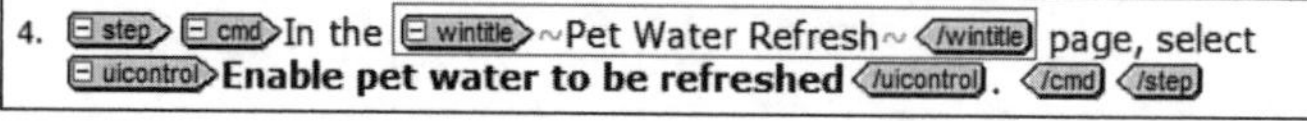

Figure 12.15 A step that correctly uses the <wintitle> element around the page called "Pet Water Refresh."

Problem: Using the <b> element instead of the <uicontrol> element for user interface controls in software documentation

Incorrect

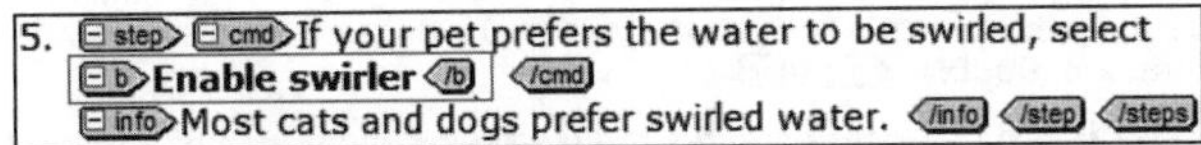

Figure 12.16 A step that incorrectly uses the <b> element around user interface control.

Correct

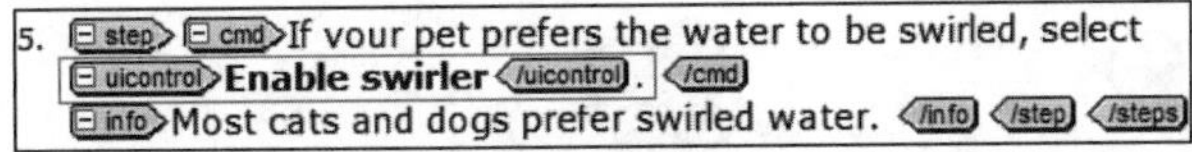

Figure 12.17 A step that correctly uses the <uicontrol> element around user interface control.

Common Problems Found in Concept Topics

In concept topics, you often find the following markup problems:

Problem: Creating long paragraphs and dense text

Break up long paragraphs (<p> element), by using unordered lists (<ul> element), definition lists (<dl> element), or add images.

Incorrect

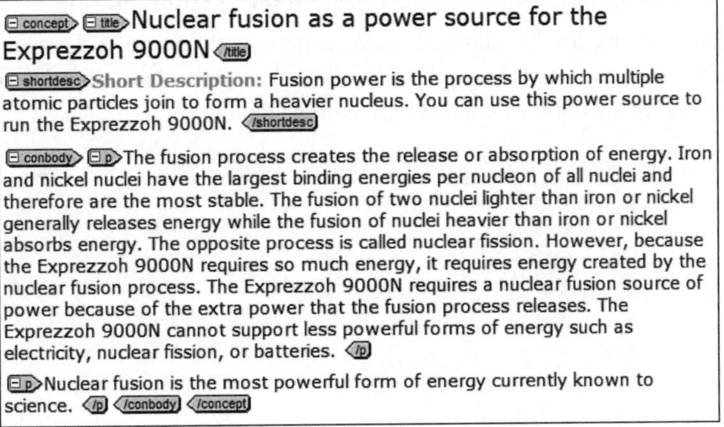

Figure 12.18 Long, dense paragraphs in a concept topic.

Correct

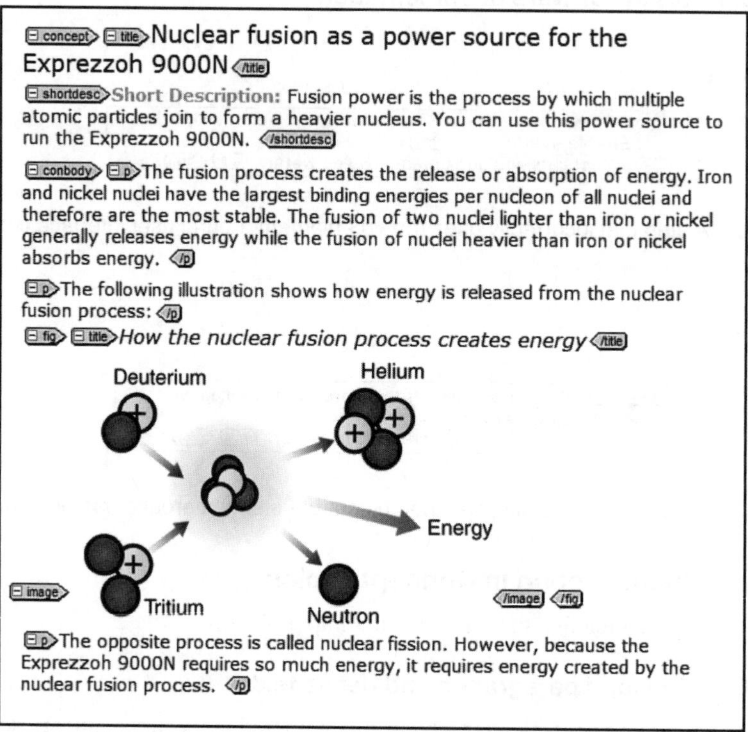

Figure 12.19 An effective concept topic that has short paragraphs and an image.

Problem: Misusing definition lists to create nested subheadings

You can insert only one level of subheadings by using <section> and <title> elements. Don't use <dl> elements to create a third level of subheadings. If you need more than two levels of headings, consider reorganizing your content into multiple topics.

Incorrect

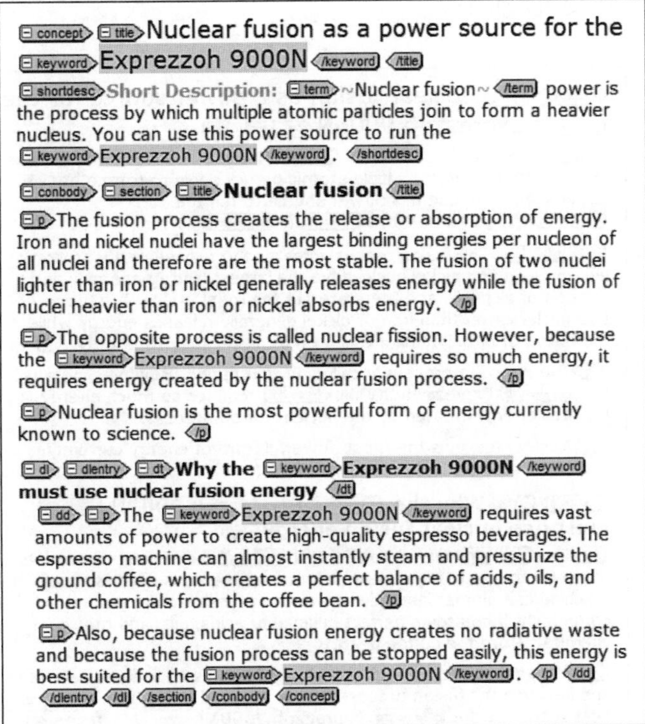

<concept><title>Nuclear fusion as a power source for the <keyword>Exprezzoh 9000N</keyword></title>

<shortdesc>Short Description: <term>~Nuclear fusion~</term> power is the process by which multiple atomic particles join to form a heavier nucleus. You can use this power source to run the <keyword>Exprezzoh 9000N</keyword>. </shortdesc>

<conbody> <section> <title>**Nuclear fusion**</title>

<p>The fusion process creates the release or absorption of energy. Iron and nickel nuclei have the largest binding energies per nucleon of all nuclei and therefore are the most stable. The fusion of two nuclei lighter than iron or nickel generally releases energy while the fusion of nuclei heavier than iron or nickel absorbs energy. </p>

<p>The opposite process is called nuclear fission. However, because the <keyword>Exprezzoh 9000N</keyword> requires so much energy, it requires energy created by the nuclear fusion process. </p>

<p>Nuclear fusion is the most powerful form of energy currently known to science. </p>

<dl> <dlentry> <dt>**Why the <keyword>Exprezzoh 9000N</keyword> must use nuclear fusion energy** </dt>

<dd> <p>The <keyword>Exprezzoh 9000N</keyword> requires vast amounts of power to create high-quality espresso beverages. The espresso machine can almost instantly steam and pressurize the ground coffee, which creates a perfect balance of acids, oils, and other chemicals from the coffee bean. </p>

<p>Also, because nuclear fusion energy creates no radiative waste and because the fusion process can be stopped easily, this energy is best suited for the <keyword>Exprezzoh 9000N</keyword>. </p> </dd> </dlentry> </dl> </section> </conbody> </concept>

Figure 12.20 A concept topic that uses a <dl> element to created artificial headings.

Correct

The corrected topic uses only one <section> element and doesn't use a <dl> element (definition list) as a third level heading.

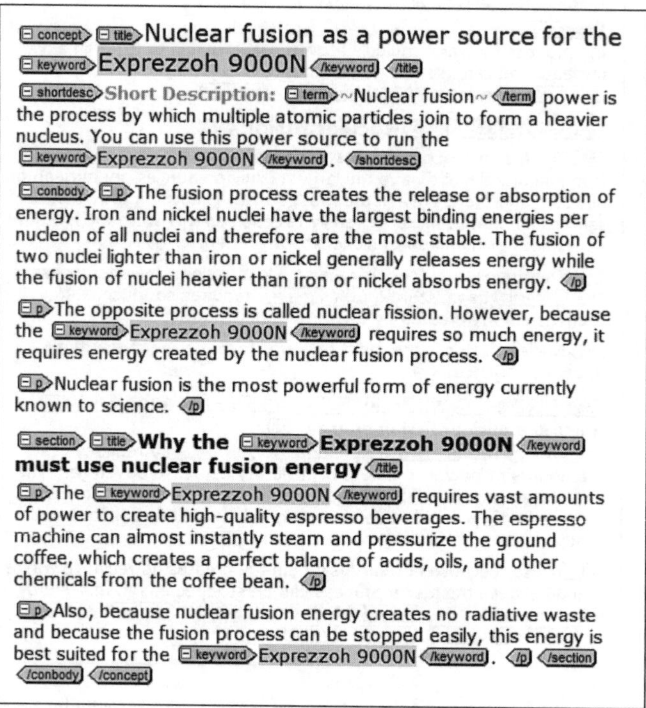

Figure 12.21 Concept topic that uses only <section> and <title> elements for headings.

Problem: Using ordered lists in concept topics to include task information. Move the task information to a task topic

Incorrect

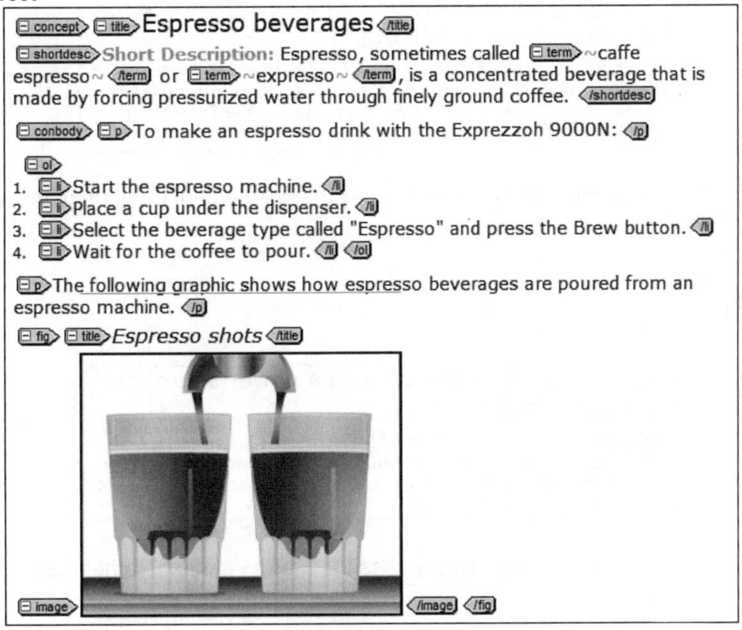

Espresso beverages

Short Description: Espresso, sometimes called ~caffe espresso~ or ~expresso~, is a concentrated beverage that is made by forcing pressurized water through finely ground coffee.

To make an espresso drink with the Exprezzoh 9000N:

1. Start the espresso machine.
2. Place a cup under the dispenser.
3. Select the beverage type called "Espresso" and press the Brew button.
4. Wait for the coffee to pour.

The following graphic shows how espresso beverages are poured from an espresso machine.

Espresso shots

Figure 12.22 A concept topic that incorrectly uses an <ol> element.

Correct

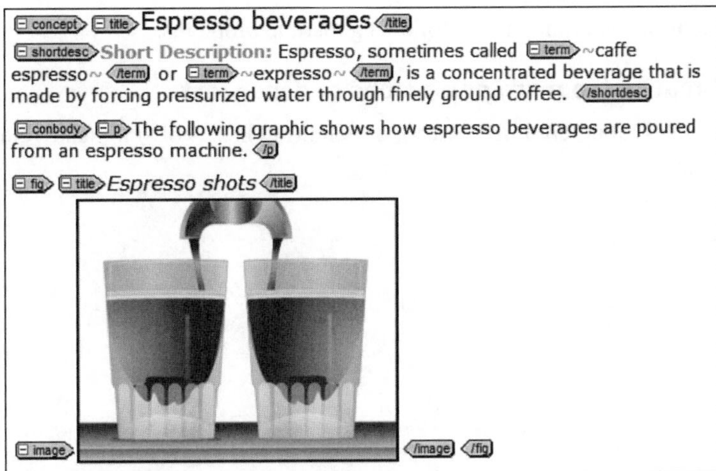
Espresso beverages
Short Description: Espresso, sometimes called ~caffe espresso~ or ~expresso~, is a concentrated beverage that is made by forcing pressurized water through finely ground coffee.
The following graphic shows how espresso beverages are poured from an espresso machine.
Espresso shots

Figure 12.23 A concept topic without an <ol> element.

Problem: Using the <i> element (italic) instead of the <term> element for new terms

Incorrect

Figure 12.24 A concept topic that incorrectly uses the <i> element rather than the <term> element for a new term.

Correct

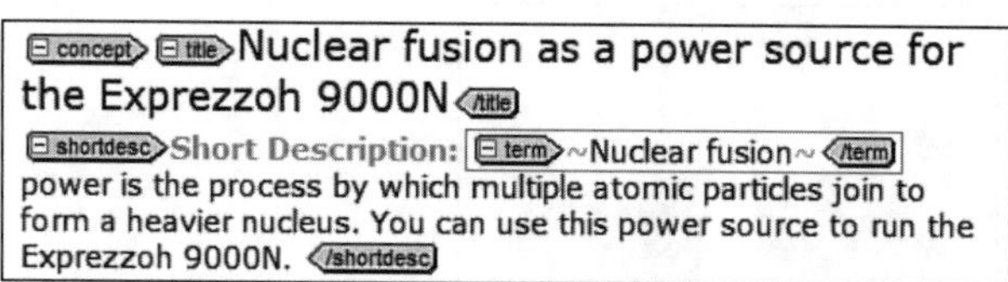

Figure 12.25 A concept topic that correctly uses the <term> element rather than the <i> element for a new term.

Common Problems Found in Reference Topics

In reference topics, you often find the following markup problems:

Problem: Creating too many paragraphs or dense text

This include tables, unordered lists (<ul>), or definition lists (<dl>) to make reference information easy to scan.

Incorrect

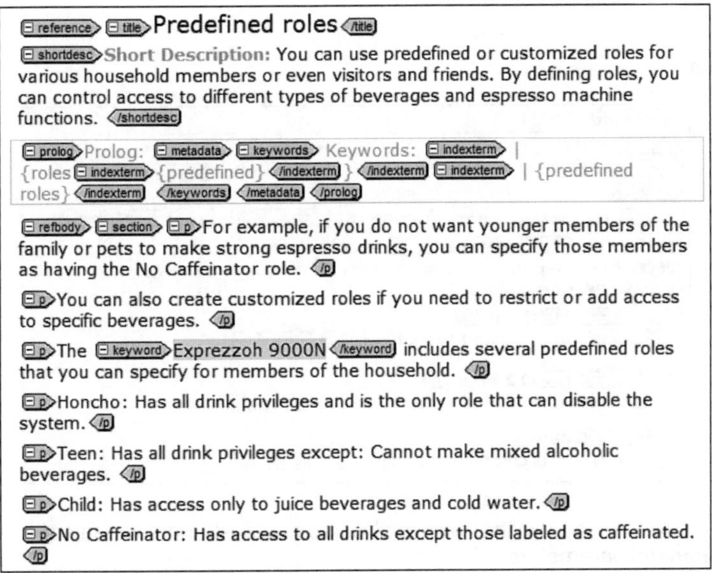

Predefined roles

Short Description: You can use predefined or customized roles for various household members or even visitors and friends. By defining roles, you can control access to different types of beverages and espresso machine functions.

Prolog: Keywords: | {roles {predefined} } | {predefined roles}

For example, if you do not want younger members of the family or pets to make strong espresso drinks, you can specify those members as having the No Caffeinator role.

You can also create customized roles if you need to restrict or add access to specific beverages.

The Exprezzoh 9000N includes several predefined roles that you can specify for members of the household.

Honcho: Has all drink privileges and is the only role that can disable the system.

Teen: Has all drink privileges except: Cannot make mixed alcoholic beverages.

Child: Has access only to juice beverages and cold water.

No Caffeinator: Has access to all drinks except those labeled as caffeinated.

Figure 12.26 A reference topic that ineffectively uses only the <p> element to organize information.

Correct

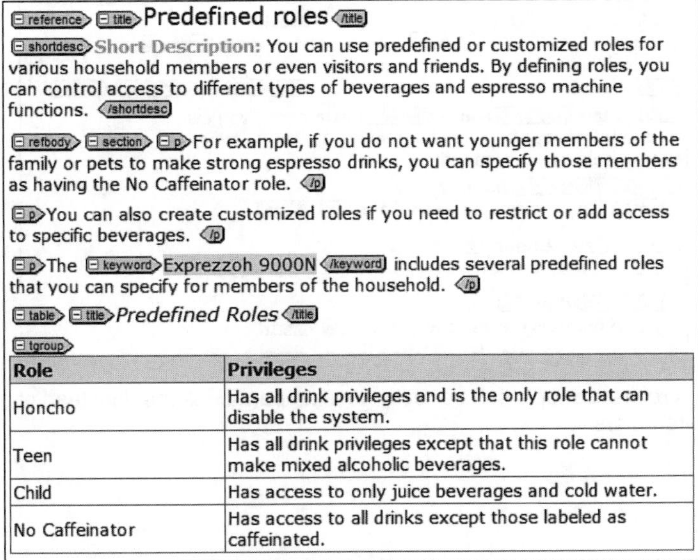

Predefined roles

Short Description: You can use predefined or customized roles for various household members or even visitors and friends. By defining roles, you can control access to different types of beverages and espresso machine functions.

For example, if you do not want younger members of the family or pets to make strong espresso drinks, you can specify those members as having the No Caffeinator role.

You can also create customized roles if you need to restrict or add access to specific beverages.

The Exprezzoh 9000N includes several predefined roles that you can specify for members of the household.

Predefined Roles

Role	Privileges
Honcho	Has all drink privileges and is the only role that can disable the system.
Teen	Has all drink privileges except that this role cannot make mixed alcoholic beverages.
Child	Has access to only juice beverages and cold water.
No Caffeinator	Has access to all drinks except those labeled as caffeinated.

Figure 12.27 A more effective reference topic that uses a <table> element to organize information.

Problem: Describing parameters or options with a definition list rather than with a parameter list (<parml>)

Incorrect

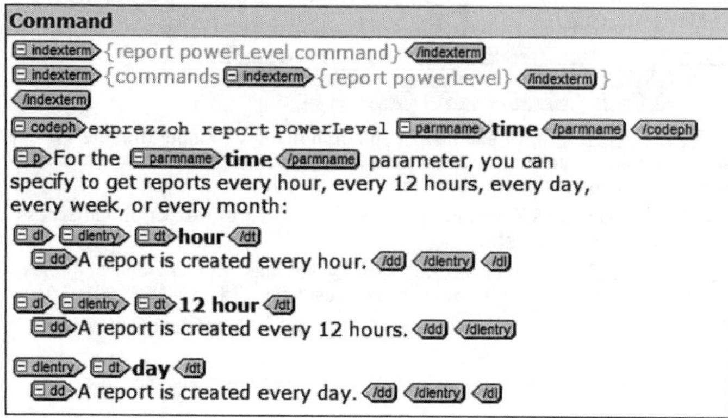

Figure 12.28 A reference topic that incorrectly uses a <dl> element instead of a <parml> element for command parameters.

Correct

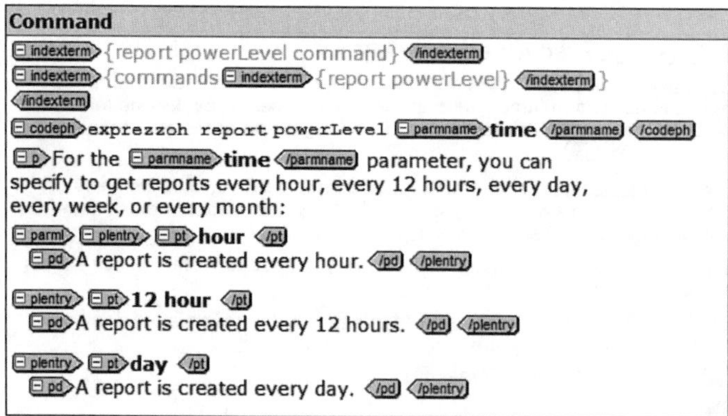

Figure 12.29 A reference topic that correctly uses a <parml> element instead of a <dl> element for command parameters.

Problem: Missing short descriptions

Even if the topic is short, you should include short descriptions in every topic. If you decide not to include short descriptions for specific types of topics, such as short API reference topics, be consistent.

Incorrect

```
<reference><title>getCode method</title>
<refbody><section><title>Syntax</title>
<codeblock>public int getCode()</codeblock>
</section></refbody></reference>
```

Figure 12.30 A reference topic that is missing a <shortdesc> element.

Correct

```
<reference><title>getCode method</title>
<shortdesc>Short Description: This method returns the status code from the
listener.</shortdesc>
<refbody><section><title>Syntax</title>
<codeblock>public int getCode()</codeblock>
</section></refbody></reference>
```

Figure 12.31 A reference topic that uses a <shortdesc> element even for a very short topic.

Common Problems Found in All Topic Types

You might see information written in the wrong topic types. For example, the content is written in a concept topic type, but the information should be in a reference topic type. Or the generic topic type is used when a task, concept, reference, or specialized topic type is a better choice.

Another problem is when you use the composite topic type instead of separate task, concept, or reference topic files. You rarely need to use the composite topic type. If you do use it, establish guidelines for when it's appropriate and use it consistently.

In all topic types, you often find the following markup problems:

Problem: Placing index elements outside the <prolog> and <metadata> elements or forgetting to insert index entries

For long reference topics, you place the index entries near the object that you want users to find. However, in most topics, place index entries in the <prolog> element.

Incorrect

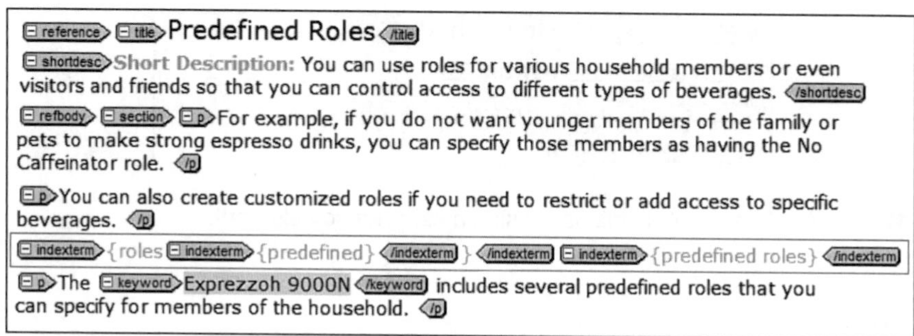

Figure 12.32 Index entries that are incorrectly placed just before the table rather than in the <prolog> element.

Correct

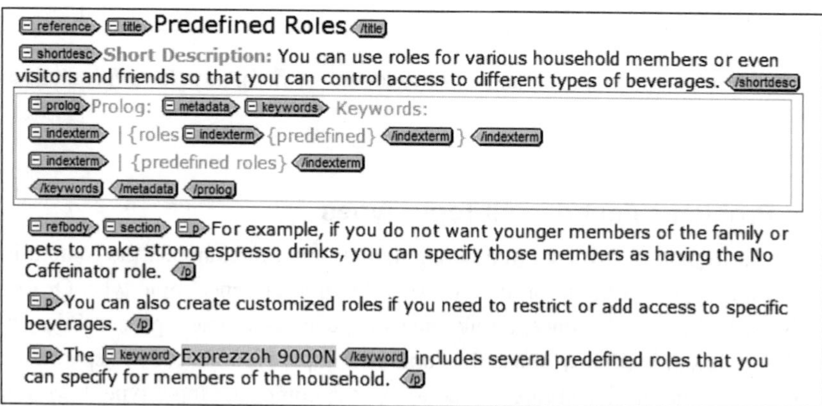

Figure 12.33 Index entries that are correctly placed in the <prolog> element.

Problem: Specifying the wrong note type or overusing the <note> element

Avoid the label "Note." Use a more specific label, such as Tip, Restriction, or Important.

Incorrect

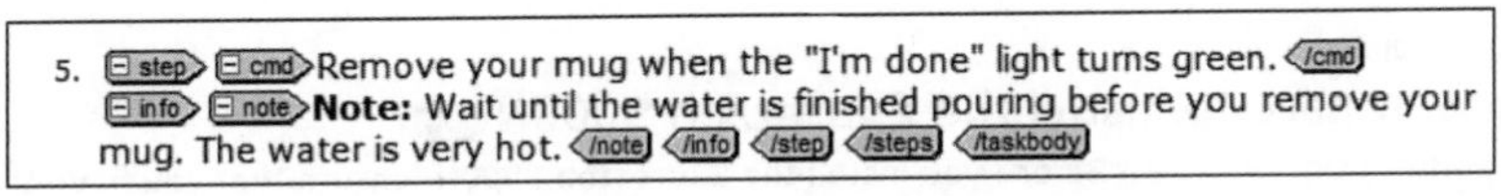

5. step cmd Remove your mug when the "I'm done" light turns green. /cmd info note **Note:** Wait until the water is finished pouring before you remove your mug. The water is very hot. /note /info /step /steps /taskbody

Figure 12.34 A <note> element that uses the generic label "Note."

Correct

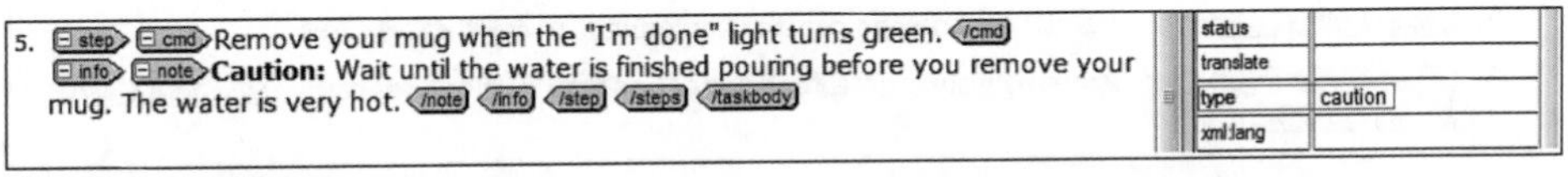

5. step cmd Remove your mug when the "I'm done" light turns green. /cmd info note **Caution:** Wait until the water is finished pouring before you remove your mug. The water is very hot. /note /info /step /steps /taskbody

status	
translate	
type	caution
xml:lang	

Figure 12.35 A <note> element that uses the more precise label of "Caution."

Problem: Duplicating links: hardcoded cross-references that use the <xref> element that duplicate automatically generated child or related links

Use hierarchical linking, relationship tables, and collection-types to control links. Avoid inline links.

Incorrect

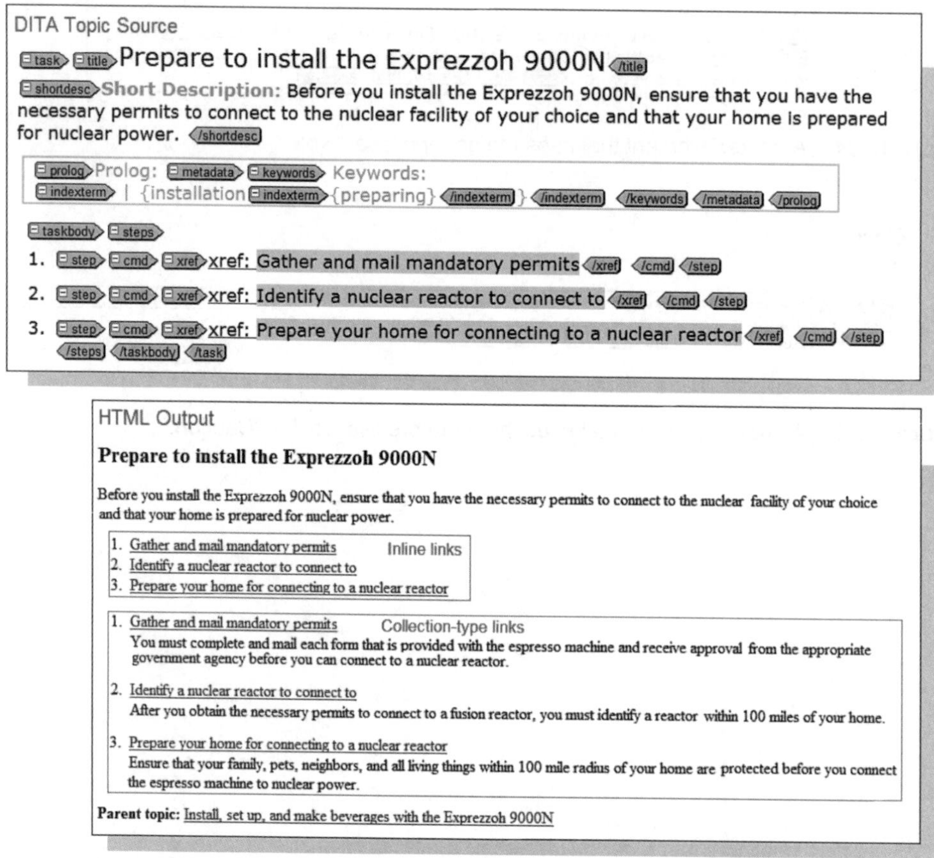

Figure 12.36 Duplicate links in HTML output created by inserting <xref> elements in the topic "Prepare to install the Exprezzoh 9000N" and using the sequence collection-type in the DITA map.

Correct

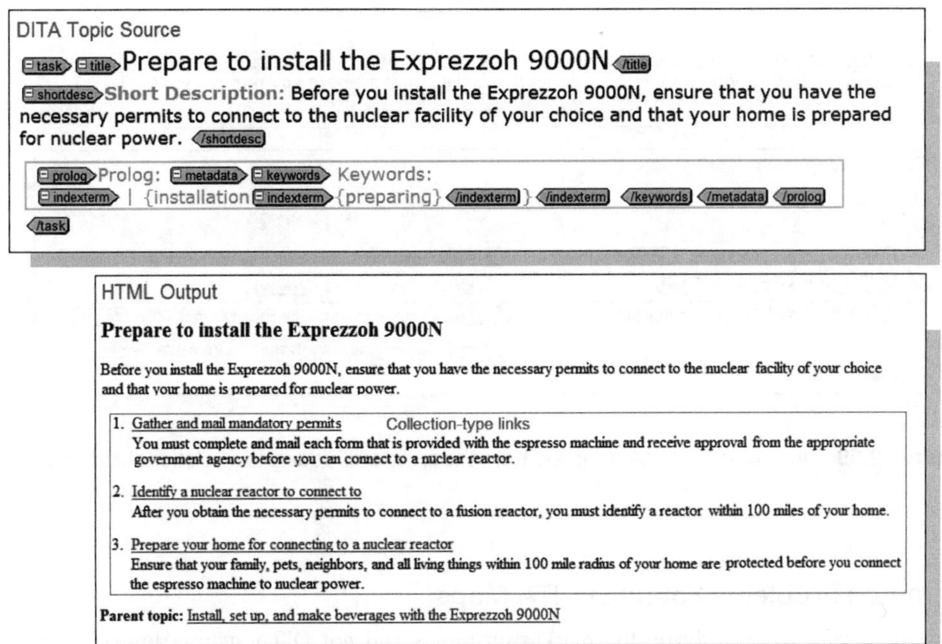

Figure 12.37 HTML output of properly constructed links in a sequence.

Problem: Adding links in the topic instead of using a relationship table

You can more easily manage links in relationship tables.

Incorrect

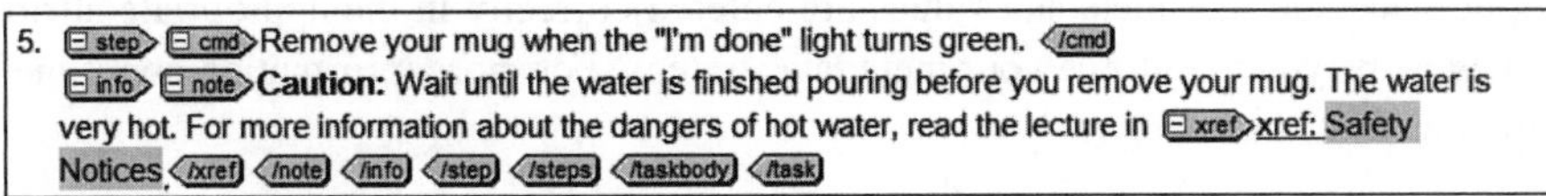

Figure 12.38 An inappropriate inline link.

Correct

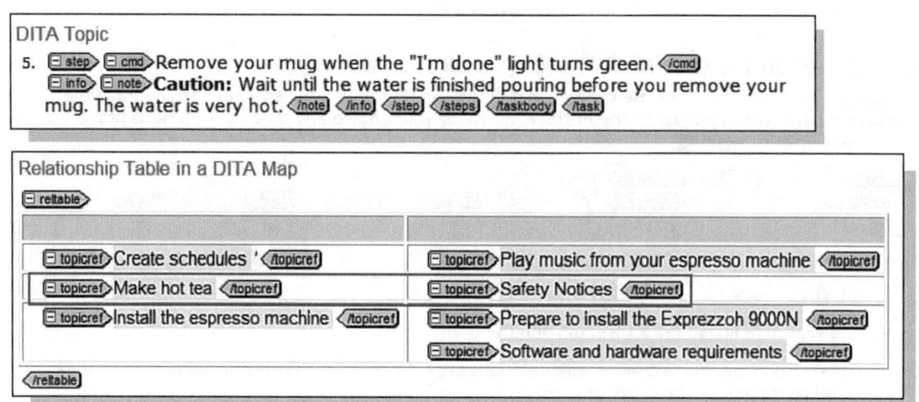

Figure 12.39 A cross-reference removed from the topic and added to a relationship table.

Common Problems Found in DITA Maps

Code reviews focus mostly on the markup in topics and not DITA maps. However, you can review the markup in DITA maps to correct problems in navigation, short descriptions, and linking. Even if you review the markup in DITA maps, be sure that you also review the output of those DITA maps. Editors typically review topic organization and linking as part of content editing (see Chapter 13, "Content Editing," for more information).

Common Problems Found in Metadata

You can also review metadata during code reviews to see whether the writers applied the metadata incorrectly or whether they applied it at all. For example, check that index entries are added in the right place and that metadata values are entered correctly in the appropriate attribute. A code review helps you to enforce your metadata strategy. For more information about metadata, see Chapter 8 "Metadata."

Step 4: Discuss Review Findings

After the code review team has reviewed the DITA files, schedule a review meeting with the writers to discuss the findings, answer any questions, and create a plan for implementing the changes. Typically, a code review meeting lasts about an hour. At the conclusion of the meeting, the expectation is that the writers will implement the code changes to all topics they own.

BEST PRACTICE Use the code review as an educational opportunity for your entire team. If several writers work on the topics, invite all those writers to the code review meeting. You might also invite other writers of other topics in the library to attend so that they can also learn.

Code reviews shouldn't turn into the Spanish Inquisition. You shouldn't make writers uncomfortable or nervous. Be positive when you suggest corrections and discuss ways to improve the DITA markup. Also, don't forget to make positive comments when writers use DITA elements correctly.

Step 5: Complete the Code Review

At the end of any code review, you should have corrected all markup problems and be able to create output files without any errors.

The code review is complete when:

- All code review participants agree on the findings.
- All code review corrections in the sample set of topics have been implemented in all other topics in the information set.

TIP Use the code review checklist that appears at the end of this chapter to help you conduct code reviews in your organization.

Code Reviews for Content Not in Topic Form

Many companies that create technical information for their products might have thousands, hundreds of thousands, or even millions of pages of documentation. Because converting all that content to DITA and organizing that content into effective topics is time consuming, you might consider conducting code reviews for content that is not in proper topic form.

To be clear, content that is in DITA but not in proper topic form can be described as:

- Topics with text that points to information just before or after those topics. Such information is sometimes called "glue" or transition text.
- Topics that serve only as introductions or conclusions and don't make sense on their own.
- Topics that depend on context in other topics to make sense.
- Topics that have misapplied elements and can't yet be divided into multiple topics.
- Topics that contain a mix of task, concept, and reference content.

If you first convert content to DITA and later reorganize your content into effective topics, you should conduct code reviews during each of these phases of the DITA conversion. After you convert content to DITA and before the code review, ask the writers to clean up the converted content as much as possible so that that content is free from transform errors. The markup in these file shouldn't cause transform errors but might not use DITA elements correctly.

You also should review metadata and processing attributes. You might have to wait months or even years before you get all your content into effective topics, so you need to ensure that metadata and processing attributes are properly set now.

When the content is not in well-formed topics, you might need to forego changing some of the DITA markup until the topics can be properly architected. For example, in DITA task topics, you shouldn't nest steps more than two levels. Unfortunately, clever folks can create nested steps to more than two levels by adding ordered or unordered lists in <info> elements. However, until the content is formed into proper topics, the markup will work, meaning that it won't produce errors, but such coding should be discouraged.

To Wrap Up

Code reviews have several advantages and can save you much rework months or years later if everyone plays by the same markup rules.

Doing regular code reviews for new writers or for writing teams that have just converted content to DITA has these benefits:

- Ensures that most or all topics contain properly applied and consistent markup, which makes your topics easy to maintain, convert, reuse, or translate
- Helps to teach new writers about DITA markup
- Reveals problems that need to be resolved, such as:
  - When to apply DITA elements to some new type of content
  - What guidelines to use when writing text for some element
  - Whether to apply a DITA element at all

To do a code review, follow these steps:

1. Schedule the review.
2. Optional: Edit the content for grammar, punctuation, and style before it gets a code review.
3. Review the DITA code.
4. Hold a meeting to review the findings with the writers.
5. Ask the writers to make changes that you discussed during the code review to all their topics.

You should schedule code reviews for all writers regardless of how long they've been using DITA. The experienced folks will have much to offer; the new folks will have much to learn; and everyone can help improve the quality of the DITA markup.

Code Review Checklist

Guideline	Description
Check task topics.	• Use <steps>, <substeps>, and <stepxmp> elements instead of ordered lists (<ol> element) in the <context> element. • Watch out for too many levels of nested steps, for example, using the <info> and <ol> elements in a <substep> element to add a third level of ordered steps. • Use the importance attribute for optional steps. • Use the <userinput> element in task steps where the user must enter content. • Use the <cmdname> element for command names in body or inline text rather than in the <step> and <cmd> elements in tasks. • Use the <wintitle> element for window, page, or panel names. • Use the <uicontrol> element for user interface controls, not the <b> element.
Check concept topics.	• Shorten long paragraphs and dense text. Include more paragraphs (<p> element) and perhaps unordered lists (<ul> element) or definition lists (<dl> element). • Don't use definition lists to create nested subheadings. You can insert only one level of subheadings by using <section> and <title> elements. Don't use <dl> elements to create a third level of subheadings. • Don't use ordered lists (<ol> element) to include task information. Move the task information to a task topic. • Use the <term> element for new terms, instead of the <i> element (italic).
Check reference topics.	• Shorten or omit long paragraphs and dense text. Include tables, unordered lists (<ul> element), or definition lists (<dl> element). • Use the <parml> element for parameters or options, not a definition list (<dl> element). • Ensure all topics, including reference topics, have short descriptions. Even if the topic is short, include short descriptions in every topic. If you decide not to include short descriptions for specific types of topics, such as short API reference topics, be consistent about when to exclude short descriptions.

Guideline	Description
Check that the correct topic type is used.	• Ensure that you use task, concept, reference topic types for the right content. • Avoid using the generic topic type if a task, concept, reference, or specialized topic type is a better choice. • Avoid using the composite topic type. You rarely need to use the composite topic type. If you do use it, establish guidelines for when it's appropriate and use it consistently.
Check that the correct metadata is applied.	• Check for missing or incorrect metadata in the <prolog> element, in the DITA map, on the root element of the topic, and on elements that should be conditionally processed. Be sure to follow your metadata strategy. • Place index elements in the <prolog> and <metadata> elements.
Check that <note> elements are used correctly.	• Specify the appropriate note type, such as Tip, Restriction, or Important. Avoid the label "Note." • Avoid using too many <note> elements. Overusing notes is like overusing exclamation points: after a while, both lose their impact.
Add comments where needed.	• Use XML comments or <draft-comment> elements. Remember that draft comments can be seen in the draft output. • Use tools to track your changes or to compare files if your authoring tool has these features.
Check other elements.	• Ensure that writers use the correct semantic elements to mark up the content. • Don't focus on how the output will appear. For example, avoid bold (<b> element) and italic (<i> element) because they are not semantic elements. • Ensure that task topics use the correct elements, such as <steps>, not elements such as ordered lists with the <ol> element. • Remove empty elements, such as extraneous paragraph (<p> element) elements. Never use empty paragraph elements to create additional line spacing. • Check for these most commonly misapplied elements such as <filepath>, <codeph>, <varname>, <apiname>, <cmdname>, <term>, <msgblock>, <note>, and <codeblock>.
Check maps and linking.	• Follow your navigation and linking guidelines that you've established for your team. • Remove duplicate links, such as hard-coded cross-references that use the <xref> element that duplicate automatically generated child or related links.

CHAPTER 13

Content Editing

Editing is essential to ensure the quality of your information. When you move to DITA, you might consider new ways to handle your traditional content editing process by editing not only the DITA markup but also the content itself in the DITA source files.

In addition to performing code reviews, you can do more common edits of the information in the DITA source to identify problems in grammar, style, punctuation, concreteness, completeness, clarity, and other areas. If you've edited content in other formats such as Microsoft Word® or SGML, you already understand the benefits of reviewing source files. When you edit in DITA, the advantages can be even greater.

The idea of editing in DITA source files rather than output files might sound a bit controversial or difficult, for example:

- Some writers might feel a sense of ownership for their files and feel uncomfortable letting other team members make changes directly in their files.
- Some writers might fear that editors will introduce errors or invalidate the content.
- Editors might think that their DITA skills are not sufficient to work in the source files.

Nevertheless, you might find that the advantages of editing content in DITA outweigh the trepidations of the team.

Traditionally, editors edit content on hardcopy or in electronic form so that they can see the documentation in the same format that users see it. Understanding the user experience is important. However, by editing in DITA, you can point out problems in the DITA markup and the content at the same time.

When editors review DITA files, they can review not only style, grammar, punctuation, and clarity, but can also evaluate navigation, linking, and DITA markup practices.

Editing the content directly in DITA files has several advantages:

- **Edit more comprehensively:** You can edit both the content and the DITA markup at the same time. Although you typically verify markup quality in code reviews, a good way to ensure that writers apply correct and consistent markup is to check the markup as you edit the content.
- **Edit iteratively:** In a fast-paced, iterative development environment, you might find it easier and faster to edit the content in DITA during each iteration of the development cycle rather than wait for writers to create the output for a complete, 200-page user guide or online help system. Writers can submit topics during each iteration instead of waiting until the end of the cycle, when it's crunch time, to assemble the complete set of information and get it edited. Editing 20 pages every few weeks is typically easier than editing 200 pages at the end of the cycle.
- **Save time and resources:** You can reduce the time and effort required to incorporate editing comments by making changes and submitting comments in the DITA source files. If you edit the content in the DITA files, writers won't need to create output files, such as PDF, submit those files, and then incorporate handwritten or electronic comments back to the DITA source.
- **Get more edits implemented:** You can ensure that more edits are addressed by making changes in the DITA source files because writers must implement comments before the topics can be published. If writers must wait to get edits at the end of a cycle, they might not have enough time to incorporate those comments before a deadline. You might waste time and money submitting editing comments that don't get addressed.

You should still to edit the output of the DITA topics several times during your cycle, but if you're sold on the advantages of editing content in the DITA source files, we recommend that you:

- Set up a process and guidelines for scheduling, submitting, returning, and tracking edits.
- Learn to use the commenting and tracking features of your DITA authoring environment.
- Edit your output to review linking and organization in addition to editing the DITA source.

Defining, Scheduling, and Submitting Content Edits

Before you edit your content in DITA, you should decide what types of edits you want to do, , when to schedule the content for editing, and how to submit the files for editing.

Defining the Types of Content Edits

Before you schedule edits, be sure that everyone on the team understands what's involved in different types of edits. Not only can you do traditional types of edits, such as copy or technical edits, but you can also do code reviews in the DITA source files.

If you haven't yet defined the types of edits for you team, you can start with the definitions in Table 13.1.

Table 13.1 Types of Content Edits

Edit Type	Description
Copy edit	In copy edits, you identify problems with grammar, style, and punctuation.
Technical edit	Technical edits are more detailed than copy edits. In technical edits, you review completeness, concreteness, retrievability, organization, and other problems beyond grammar, style, and punctuation. You also assess the logic of the topic organization and whether the information is task oriented.
Organizational edit	In organizational edits, sometimes called structural or developmental edits, you review an outline of the organization of the information. Often, writers ask for an organizational edit before they've added much content to their topics. Viewing topics in a DITA map is one way to do an organizational edit.
Code review	In a code review, you identify problems with DITA markup and structure. See "Code Reviews" in Chapter 12 for more information about reviewing DITA markup.

For more information about editing content for grammar, punctuation, and consistency, writing for international audiences, and style, see *The IBM Style Guide: Conventions for Writers and Editors* by DeRespinis et al.

For more information about editing for clarity, task orientation, organization, and retrievability, see *Developing Quality Technical Information* by Hargis et al.

Scheduling the Edits

The question of when to edit is a bit complicated because DITA topics are discrete units of information that can be independent of other content. For example, some teams might find it easier to schedule one edit for the entire installation guide rather than schedule several shorter reviews to edit only a few topics at a time.

Also, when and how you schedule edits depends on the development processes for your product and what types of edits you need to do. Your processes, schedules, and team dynamics can vary, but most technical product teams follow either an iterative or agile process, or a traditional waterfall process:

- The iterative process typically requires a fast and constant flow of product development. The team develops specific product features and creates the documentation for those features about every 2–6 weeks.
- The waterfall process can be just as demanding, but the development cycles tend to be longer, and you might need to write and edit larger portions of documentation at key milestones in the cycle.

Editing in Iterative Development Environments

In an iterative development team, writers often create a small number of topics to document the features that are developed during each iteration, or sprint. In these teams, an editor might review individual topics rather than larger sets of information.

In the early iterations of an iterative project, writers might have only a few disparate topics and not yet have DITA maps. As the development cycle progresses, the writers create more content, so editing the organization, linking, and completeness of an entire DITA map and its topics can be done only later in the cycle.

If you write and edit in an iterative product development cycle, we recommend that you:

- Create an information model or outline of the topic organization before writers create too much content. Then, review the organization of that model to ensure that it's logical and complete.
- Do technical edits of topics in DITA during each iteration, or sprint, so that you keep up with the product development schedule. Editing your content in DITA can help you keep up with the fast pace of iterative development.
- Do organizational edits of the DITA maps and output after the first several iterations to ensure that writers are following the information model.
- Schedule a final a edit near the end of the cycle to look for problems in architecture and linking in the output that you might've missed in previous topic and DITA map edits. In this final edit, you shouldn't find serious problems if you've been editing during each iteration.

Editing in Waterfall Development Environments

In a traditional waterfall cycle where the product development is not contained in well-defined iterations, writers might prefer to send larger numbers of topics or DITA maps and their topics for editing at designated milestones throughout the cycle, such as preparing for a beta or a shipment to a translation center.

However, you can follow a schedule and process similar to the iterative editing schedule even if you work in a waterfall development environment.

Submitting Content for Editing

How your writers submit DITA source files for an edit varies depending on your tools and processes. Nevertheless, you might use the following methods and tools for submitting files for review:

- Design the workflow in your content management system to route topics and DITA maps to designated editors.
- Use a collaboration tool, such as a Lotus® Notes® database or community, Microsoft SharePoint®, or a wiki, to submit and track edit requests.

WATCH OUT Don't use email to distribute files. Email lacks version control, audit history, and other useful review features. Using email for reviews might also violate auditing processes that are mandatory for your company.

Whether you work in an iterative or waterfall development environment, remember that you can submit one or more topics per edit. One advantage of topic-based writing is that you create self-contained, independent topics. By writing in topics, you can more easily submit individual topics or small groups of topics for editing.

The editor doesn't always need to see all the other related topics to evaluate the quality of a topic. The editor's job should be to evaluate whether a topic is properly self-contained and that the content is written in the correct topic type.

In addition to submitting one topic at a time for an edit, writers can also submit a small group of related topics. For example, a writer might be working on a user manual that describes how to assemble a motorcycle engine. Instead of waiting until the end of the product development cycle to submit the entire manual or long chapter, the writer can submit individual sets of topics that describe assembling the carburetor or adjusting the valve clearance.

Providing Editorial Feedback

Use the comment elements in DITA and the editing features of your XML authoring tool to provide feedback to writers in the DITA files.

Inserting Draft Comments

Use <draft-comment> elements to include comments or questions about content. More specifically, use the <draft-comment> element to:

- Ask a writer to clarify a statement
- Explain why you are suggesting certain edits
- Ask a writer to add more information to correct content gaps
- Suggest ways to improve topic flow or organization
- Remind writers about the company style guidelines

Figure 13.1 shows a draft comment from the editor that asks the writer to clarify a statement.

You can also use the author, time, and disposition attributes of the <draft-comment> element to include more information about the source of the draft comment. This information is useful for determining who entered the comment, when the comment was inserted, and whether the comment was addressed.

Figure 13.2 shows attribute values for a <draft-comment> element.

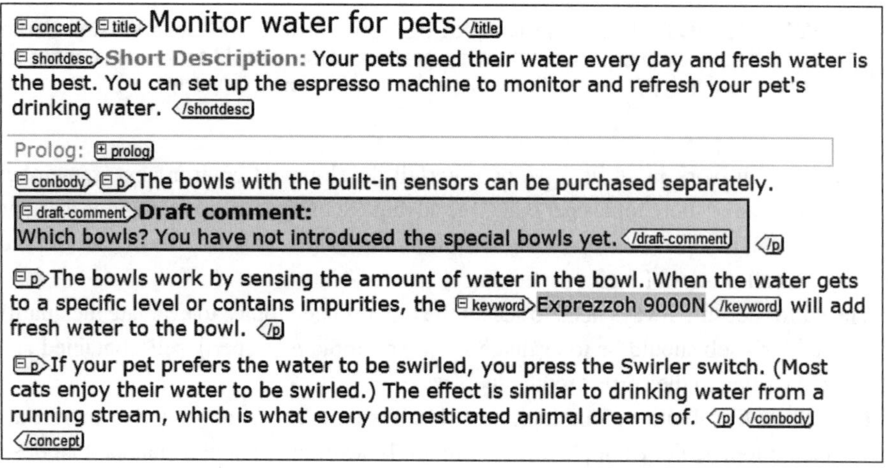

Figure 13.1 A <draft-comment> element that's used to provide feedback to writers.

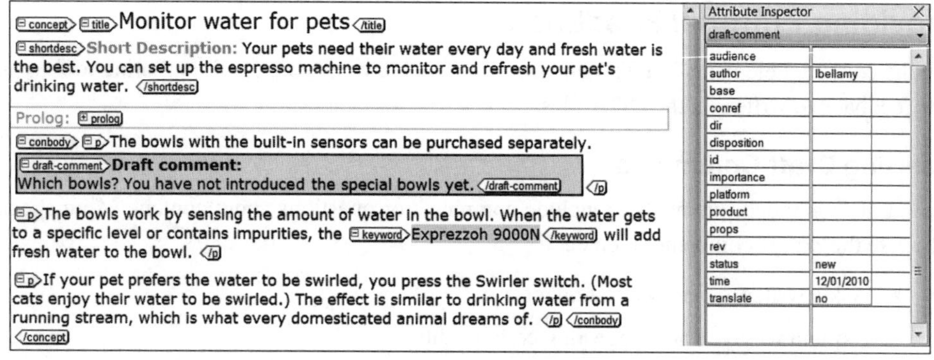

Figure 13.2 Attributes set on a <draft-comment> element.

You can modify the style sheet of your XML authoring tool to display attribute values, such as author and time, of the <draft-comment> element in topic text, as shown in Figure 13.3.

The <draft-comment> element was designed to appear in draft output only. For example, you might want to include draft comments in your draft output for technical reviewers or peer writers.

Ensure that all <draft-comment> elements are removed from the DITA source files before the content is released to customers. Unless you remove draft comments manually, you must modify the options in the output processing to prevent draft comments from being shown in the final version of the output.

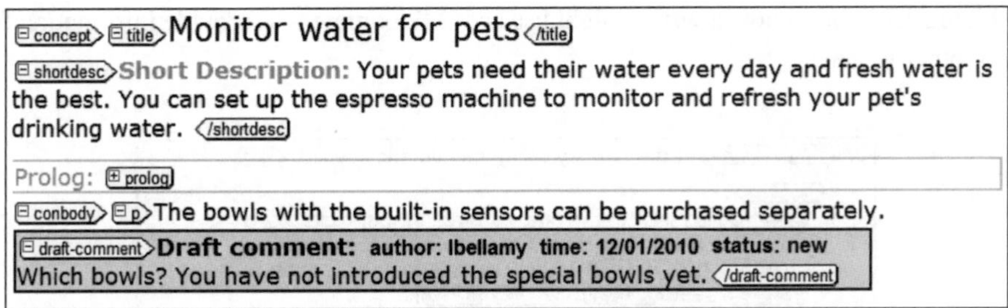

Figure 13.3 Attribute values that are shown in the draft comment text.

Decide with your team how to handle <draft-comment> elements. For example, your process might be that writers simply address the comments and delete the <draft-comment> element, or that the writers update the <draft-comment> element to say that the change was incorporated.

TIP When you do include <draft-comment> elements in the output, evaluate how effective the highlighting is. Draft comments should be easy to recognize and distinct from other text for your reviewers. Modify your style sheet to highlight content included in the <draft-comment> element for easy identification.

Inserting XML Comments

Use XML comments to include comments in DITA maps and topics that should persist across releases. XML comments are a standard way to include comments that are suppressed from the output.

An XML comment is not a DITA element, but you can insert XML comments in most places in a topic or DITA map. The following example is an XML comment:

```
<!-- This is an XML comment. -->
```

Because text in XML comments is suppressed from the output, these comments don't need to be deleted before the output of the content is released to customers.

Use XML comments to:

- Describe content that might be an exception to your style or markup guidelines.
- Provide instructions and clarification to other writers or editors.
- Record action items for future releases.
- Add text that you want to maintain but that you don't want to expose in the output.

Figure 13.4 shows how a writer might use an XML comment to provide information for an editor or other writer.

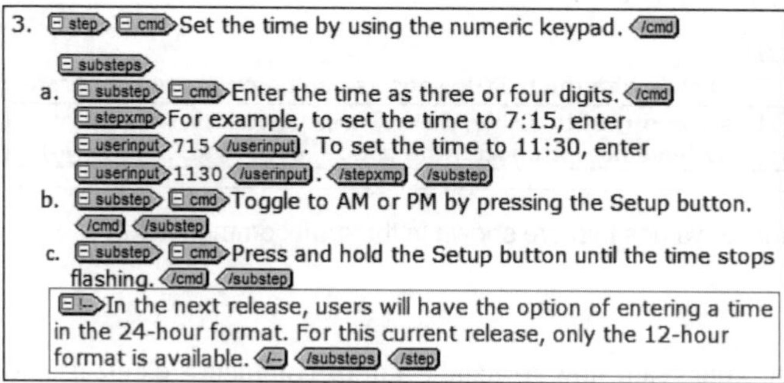

Figure 13.4 Information in an XML comment.

If you provide DITA source files to customers, third-party vendors, or business partners, consider removing XML comments from your files because you could be exposing sensitive information about your company or organization to external users.

Tracking Changes

Most XML editors have a track changes feature that shows insertions and deletions in the DITA files. Use the track changes feature in your XML editor to highlight changes that you make in the DITA files.

Writers often say that they don't have enough time to address edits. By using a track changes tool, you can make changes in the DITA files, and writers can accept or reject those edits quickly. By using a track changes tool, you can improve your editing process by:

- Visually highlighting the changes
- Including the suggested changes directly into the DITA file rather than on hardcopy
- Reducing the effort for writers to evaluate and incorporate edits
- Improving the chances that edits will be addressed
- Providing a history of edits and whether the writer accepted them

When you return the edited files, writers can accept or reject the suggested changes. If writers accept the change, the change is incorporated into the DITA file and the highlighting is removed. If writers reject the change, the change is removed and any deleted content is restored in the file.

For example, Figure 13.5 shows edits that suggest grammatical and organizational improvements. The editor suggests that prerequisite information be moved from the <context> and <step> elements to the <prereq> element. To incorporate these changes into the DITA topic, the writer simply accepts the insertions and deletions.

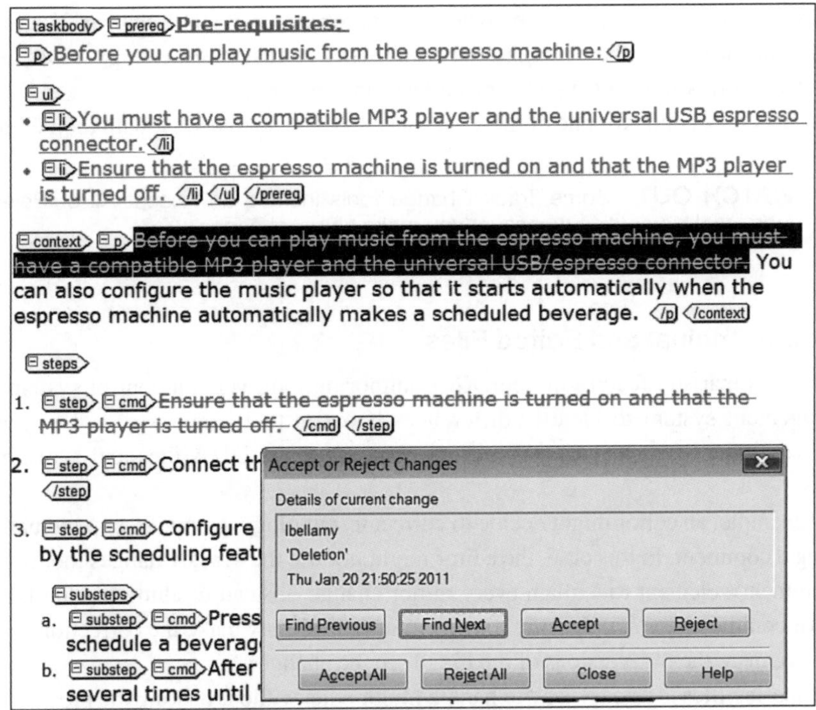

Figure 13.5 Edits that are made by using a track changes tool.

WATCH OUT All changes to the DITA files might not be tracked. In DITA map files, most changes are not tracked. In DITA topics, replacing one DITA element with another element doesn't show in Track Changes. Review the Track Changes feature of your XML authoring tool to understand which actions are tracked.

In XMetaL, changes to markup in DITA topics are not tracked. For example, if you change an element, but don't replace the text, the change might not be highlighted. You can make the change, but the writer might not be aware of the problem that you corrected. Instead of changing an element, delete the incorrect element and insert the appropriate element, as shown in Figure 13.6.

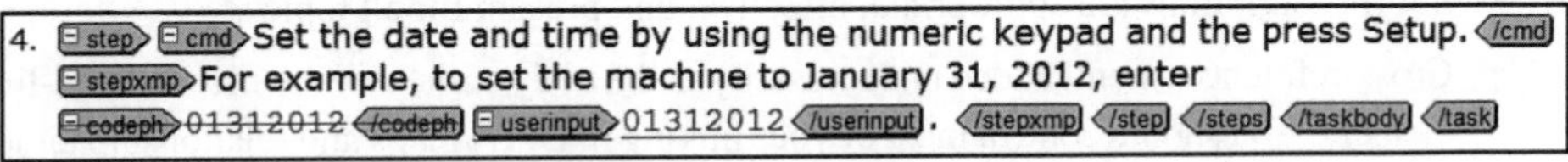

Figure 13.6 An element with a strikethrough followed by the correct element.

Establish guidelines that describe when it's appropriate to reject a change. You might discover legitimate instances where a suggested change isn't correct, or the change would reduce the quality of the information or change the meaning of the content.

If writers reject edits for one of these reasons, they should discuss it with their editor.

WATCH OUT Some Track Change tools include an Accept All or Reject All function. Use caution whenever you make global changes.

Comparing Original and Edited Files

Use the file comparison feature in your XML authoring tool, version control system, or content management system to identify differences between two versions of a file. Comparing files is handy when you want to make minor changes in the DITA files and avoid including comments.

For example, an editor might decide to correct terminology, grammar, or punctuation without leaving a comment. In this case, the editor might not use the Track Changes tool or include a <draft-comment> element to explain every minor change. Instead of cluttering the DITA topic with lots of comments for simple corrections, editors can simply make the correction and ask the writers to compare the two versions of the file to see the changes.

Comparing files is an easy way to handle minor edits without making the process too arduous for editors and writers.

Editing the Content in DITA Topics and Maps

Although you should use code reviews to identify misuse or neglect of DITA elements, you can't always separate the quality of the content in the DITA element from the DITA element itself. For example, editors must often ensure that the writer has not only used the correct element, but has also created the right content for that element based on company style guidelines.

Editing content in DITA helps you ensure that writers use the correct DITA elements and create the correct content for those elements.

Editing DITA Topics

When you edit DITA topics, you should review both the content and the DITA markup. In addition to editing the content for clarity, accuracy, concreteness, and other content quality characteristics, you should review the DITA elements and their content to ensure that:

- Short descriptions are appropriate, effective, and present for all topics.
- Cross-references aren't added unnecessarily in the body of the topic or in a <related-links> element. You should control most of your linking in DITA maps and relationship tables.
- The correct topic type is used for the content. For example, ensure that the writer hasn't added a procedure with an ordered list (<ol> element) in a concept or reference topic.

- Task, concept, and reference information isn't mixed in one topic. Remember to separate this information into separate topics.
- Writers are following style guidelines for tables, figures, lists, task steps, and other items that require specific DITA elements and content, such as <title> elements on tables or step introductions in the <context> or <stepsection> element. (The <stepsection> element is available only in DITA 1.2.)

TIP If you have a difficult time ignoring DITA markup when you are focusing on text, change the file view in your XML authoring tool to hide the markup.

For a more comprehensive list of markup problems in the DITA topics, see Chapter 12, "DITA Code Editing."

For more information about how to create clear, complete, retrievable, and task-oriented information, see *Developing Quality Technical Information* by Hargis et al.

Editing DITA Maps

Editing DITA maps and relationship tables is just as important as editing topics. When you edit DITA maps and relationship tables, ensure that:

- Short descriptions in the DITA map follow the same guidelines for short descriptions in topics.
- DITA maps are complete and well organized:
  - Topics are logically nested.
  - The topic organization makes sense to users of the information.
  - The topic organization is consistent in an information set. For example, if you've decided to structure your information by creating task topics first with concepts and reference topics nested under those task topics, ensure that all the topics in the map follow the same organizing principle.
- Linking is correct. You'll need to review both the DITA map and the output to verify that linking is correct.
  - Links are going to and from the appropriate topics. For example, a task topic should typically be linked to relevant concept or reference topics.
  - The number of links is appropriate. Be sure the writer didn't create an excessive number of links.
  - Most topics include at least one link other than a parent topic link so that the user isn't stranded in a topic.
  - Linking attributes are set correctly for local, peer, and external links.

- Most linking is handled by hierarchical linking in the DITA map or by relationship tables and not by <xref> elements inserted in the topics.
- Parent topics use the correct collection type. For example, set the collection-type attribute on the <topicref> element to "sequence" for a group of nested task topics so that "Previous topic" and "Next topic" links appear in the output.

Editing the Output

Editing DITA files directly can save you time: You can review both the DITA markup and the content at the same time. However, you also might want to edit the output to review the content as the customer sees it.

Metadata, linking, topic organization in DITA maps, and attribute values on some DITA elements can be difficult to "see" when you're viewing just the DITA topic. For example, the collection-type attribute is a single value on a <topicref> element in the DITA map, but the value of that attribute creates linking and navigation for an entire set of topics. Unless you're an experienced DITA user, you can't easily envision the organization of a set of topics and automatically generated links for those topics.

To understand the experience of your users when they read the documentation, you should also review the information in the same way that your users will see it. You can review the DITA output in one of the following ways:

- **Build output regularly.** Build the output of the topic or DITA maps to see what the HTML or PDF output will look like.
- **Review existing builds.** If your company has a continuous or automated build machine that displays the output of the content on an internal website, use this site for reviewing the organization and links for the topics submitted for editing.
- **Request both the DITA source file and its output.** When writers submit one or more DITA topics for editing, also ask them to include the output of the DITA map that contains the topic.

To Wrap Up

If you've decided to adopt DITA, you must ensure that you use DITA elements, topics, DITA maps, and linking correctly. Although editing in DITA can save time and resources, adopting DITA hasn't made the work of editing the content magically disappear.

Editors need to continue to identify and eliminate problems with accuracy, clarity, concreteness, completeness, grammar, organization, punctuation, retrievability, style, task-orientation, and topic-based writing.

Fortunately, you can edit both the DITA markup and the content at the same time. Editing content in the DITA source files has several advantages:

- **You're more likely to get more edits implemented, and you can save time.** By making changes in the DITA files by using comments or track changes tools, writers can't ignore the edits. Editors can save time because they don't have to mark up output files, and the writers don't have to transfer the edits back in the DITA source files.
- **Your edits are more comprehensive.** You can edit both the DITA markup and the content in the same edit.
- **You can keep up with iterative development cycles.** Because writers need to create topics for each iteration, you should also edit those topics in the same iteration so that the documentation is done when the product development is done. Iterative editing can save you from having to fix serious problems late in the cycle.

Editing content takes more than just a good editor. To get those edits done, you need to establish a process and take advantage of tools:

- Decide which types of edits to do and when to do them.
- Set up the tools and processes to submit and track edit requests.
- Decide how to provide feedback in the DITA files.

Whichever tools or processes you follow, ensure that you provide the highest quality information to your users by thoroughly reviewing both the DITA source files and the output.

Content Editing Checklist

Guideline or Decision	Description
Decide which types of edits you need to do and when.	You should schedule organizational, technical, copy edits, and DITA code reviews. Schedule edits according to your product development cycle. However, whichever development process you follow, you should edit small groups of topics as writers finish them rather than wait until the end of the cycle and edit a 200-page manual or help system.
Decide how writers should submit content for review.	Set up a process and determine what tools you want to use to process edits. For example, use content management systems or collaboration tools to submit and track edit requests.

Guideline or Decision	Description
Decide on conventions for providing feedback.	Your options will depend on your XML editor and how you want to provide feedback, but most teams can use one or more of the following methods or tools: • Use XML comments and the <draft-comment> element to provide suggestions, corrections, and instructions. • Enable the Track Changes tool in your authoring tool to highlight insertions and deletions. • Compare original files to edited files by using a file comparison tool.
Edit the topic content and markup.	When you edit content and markup in topics, ensure that: • Short descriptions are effective. • There aren't too many inline links created with the <xref> element. Most linking should be handled by relationship tables or by collection types. • The correct topic types are used and that task, concept, and reference information is separated into separate topics. • Writers are following style guidelines for tables, figures, lists, task steps, and other items.
Edit the DITA map and relationship tables.	Similar to editing topics, in DITA maps you should check short descriptions and metadata. Also, you'll need to check: • Topic organization in DITA maps • Linking that's set by the collection-type attribute and the linking set in the relationship table
Edit the DITA output.	Be sure to review the output, such as HTML or PDF. You can often spot problems by looking at the output that you won't catch when you review only the DITA source files. Also, build the output often and regularly to catch serious problems early in the cycle. For example, you don't want to discover at the last minute that a number of topics were created with the wrong topic type, or that some topics have duplicate links.

Index

D

S

U

V

W

X

Your purchase of ***DITA Best Practices*** includes access to a free online edition for 120 days through the Safari Books Online subscription service. Nearly every IBM Press book is available online through Safari Books Online, along with more than 5,000 other technical books and videos from publishers such as Addison-Wesley Professional, Cisco Press, Exam Cram, O'Reilly, Prentice Hall, Que, and Sams.

SAFARI BOOKS ONLINE allows you to search for a specific answer, cut and paste code, download chapters, and stay current with emerging technologies.

Activate your FREE Online Edition at www.informit.com/safarifree

STEP 1: Enter the coupon code: PTMFHFH.

STEP 2: New Safari users, complete the brief registration form. Safari subscribers, just log in.

If you have difficulty registering on Safari or accessing the online edition, please e-mail customer-service@safaribooksonline.com